DATE DUE			

The Laboratory Recorder

LABORATORY INSTRUMENTATION AND TECHNIQUES

Series Editor: ***Galen W. Ewing***
Seton Hall University

Volume 1: ***The Laboratory Recorder***
By Galen W. Ewing and Harry A. Ashworth • 1974

In preparation

Volume 2: ***Introduction to Nuclear Radiation Detectors***
By P. J. Ouseph

A Continuation Order Plan is available for this series. A continuation order will bring delivery of each new volume immediately upon publication. Volumes are billed only upon actual shipment. For further information please contact the publisher.

The Laboratory Recorder

Galen W. Ewing

and

Harry A. Ashworth

Seton Hall University
South Orange, New Jersey

PLENUM PRESS • NEW YORK AND LONDON

Library of Congress Cataloging in Publication Data

Ewing, Galen Wood, 1914-
The laboratory recorder.

(Laboratory instrumentation and techniques; v. 1)
Includes index.
1. Recording instruments. I. Ashworth, Harry A., 1943- joint author. II. Title.
TK393.E94 502′.8 74-22364
ISBN 0-306-35301-6

A Division of Plenum Publishing Corporation
227 West 17th Street, New York, N.Y. 10011

United Kingdom edition published by Plenum Press, London
A Division of Plenum Publishing Company, Ltd.
4a Lower John Street, London WIR 3PD, England

Printed in the United States of America

Contents

1

Introduction

One of the most universal functions of any scientific or engineering laboratory is the gathering of data to provide answers to immediate questions or information to be filed for future reference. Such data gathering may be achieved in various ways, depending on the nature and quantity of the information. The most prevalent of such data gathering methods is undoubtedly analog recording.

Electrical analog recorders are available in a variety of sizes, speeds, sensitivities, and prices. They are suitable for recording any signal which is in, or can be converted to, electrical form. These recorders are found in every modern laboratory. Without them the importance of many functional relations would be missed altogether. How could one adequately diagnose a heart ailment without a cardiographic recorder, or obtain infrared or magnetic resonance spectra on any practicable basis without a strip-chart recorder?

True, various curves that are now traced automatically with a recorder can be plotted manually from point-by-point measurements. This procedure, however, is not only time-consuming, but may cause valid bits of information to be overlooked entirely, simply because the points were taken too far apart.

Another factor favoring the use of recorders is the ability to pinpoint faulty operation of the data-gathering system. Artifacts that might not be observable at all in point-by-point observations will often be readily identifiable on a recording. Asymmetry of a peaked curve, for example, is only clearly evident in a recording.

The objectives of this book are to present the principles on which recorders are based, and to describe the implementation of these principles. Such a treatment, it is hoped, will make the selection of a recorder on a rational basis much easier and will assist the user to obtain the best results from the equipment at hand.

The instruments that are described serve as examples only, and their mention is not to be considered as an endorsement. No treatment is given of magnetic tape recorders, nor of digital printers.

Overview. In this section, by way of orientation, some of the main features of laboratory recorders are outlined. Most of these are treated in more detail in subsequent chapters.

Recorders can be categorized in many ways. They may be capable of plotting one variable against another (X–Y recorders) or only against time (Y–T or simply "strip-chart" recorders). They may be equipped to plot a single variable or two or more simultaneously (one-, two-, or multiple-pen recorders), or they may be designed to plot several variables consecutively as a series of points.

Perhaps most fundamentally, recorders are classed by the kind of signals they will accept. Thus we have millivolt recorders, recording ammeters and wattmeters, either ac or dc, and many others. Frequently the scale is designated in nonelectrical units, such as pH or degrees Celsius, depending on specific transducers. The dc millivolt or microampere recorder is the most versatile, since input circuitry can adapt it to almost any other application.

The electrical or electronic mechanism for positioning the pen may be of varying degrees of complexity, from a simple moving-coil meter with its pointer extended to hold a stylus or pen, to a self-balancing servo system. To increase flexibility, peripheral circuits for zero suppression and range selection are often included.

As all too well known to many users, the pen-and-ink system may be the weakest link in an otherwise excellent recorder. To increase the reliability of this feature, designers have tried all sorts of devices for transferring information from pen to paper. The most common continues to be the capillary pen delivering liquid ink. Also in use are ball-point pens, heated styli, and electrical recording through sensitized paper, each with many modifications. In some instruments, writing is accomplished by a moving light beam impinging on photographic paper.

Another classification is based on the form of the paper employed. This may be supplied in continuous rolls or in separate sheets. The paper may be plain graph paper or it may be preprinted with coordinate scales for a particular instrumental application. Fan-folded paper can often be substituted for roll paper with perhaps greater convenience.

2

Deflection Recorders

The most straightforward type of recorder depends upon deflection of the pen by a conventional meter movement. The signal to be measured appears as a current flowing through a coil of wire pivoted or suspended between the poles of a magnet. The interaction of the magnetic field produced by the current with the static field tends to produce circular motion. This motion must be counteracted by some restoring force which is usually supplied by a pair of hairsprings, one above and one below the coil, that also serve to make electrical contacts. If the coil is suspended rather than pivoted, the restoring force must be supplied through torsion of the suspension. Rigidly attached to the coil is a pen or stylus which leaves a trace on a moving strip of paper. Figure 2-1 suggests how this may be done.

This mechanism is inexpensive, and capable of fairly high speed of response. It suffers, however, from two basic difficulties. First, the power necessary to move the pen must be supplied by the signal itself, and in the more sensitive ranges, this power is far from negligible. The second difficulty is rather less serious: The trace produced by the pen in this simple arrangement is inherently curved, and the chart paper must be printed with a corresponding curvilinear Y coordinate, as visible in Figure 2-1. Let us consider first the latter of these difficulties.

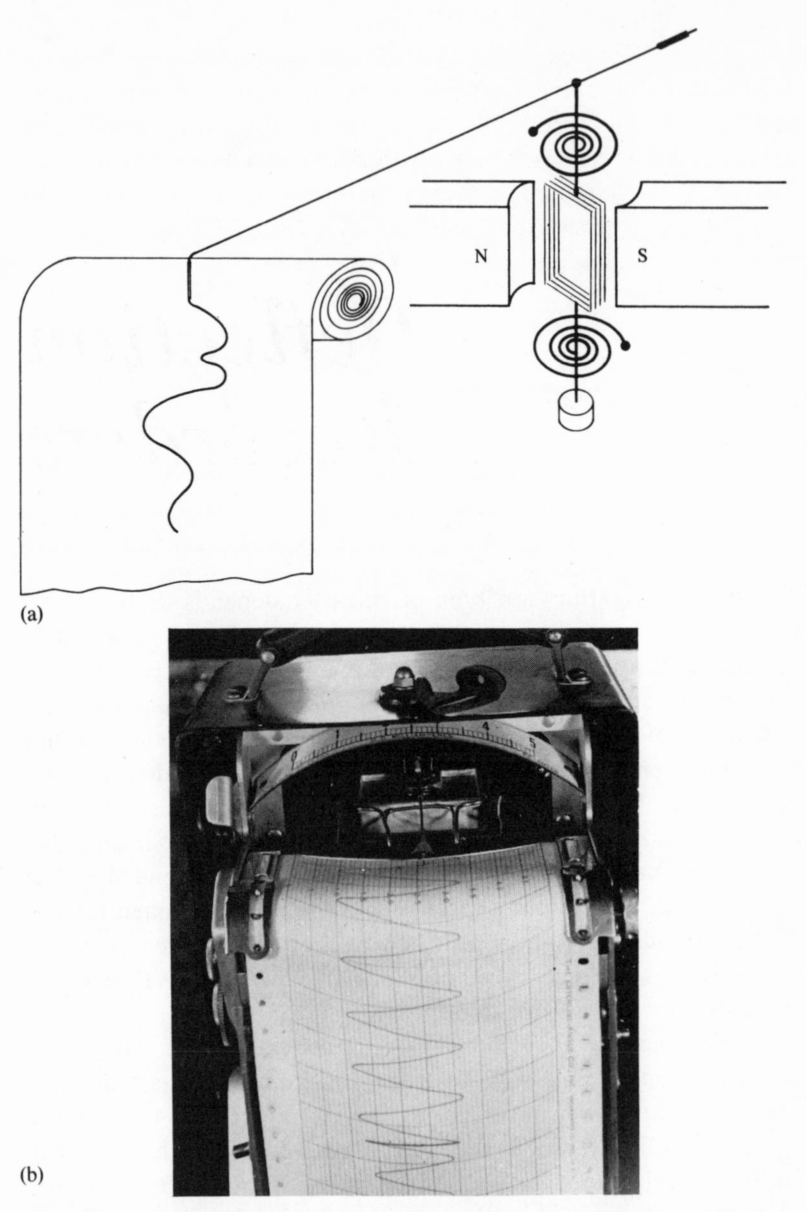

Figure 2-1. (a) A coil carrying the signal current is suspended from hairsprings between the poles of a permanent magnet. The pen is rigidly attached to the coil. (b) An Esterline-Angus deflection recorder using the mechanism described in (a).

It must be recognized that curved coordinates are entirely satisfactory for some purposes. The recording of atmospheric temperature and pressure, for example, does not suffer; the values of the variables at any desired time can be read with ease. However, for a record in which the shape of a waveform is of importance, or where the area beneath a trace is to be estimated, curved coordinates become more than an inconvenience. Figure 2-2 shows examples.

LINEARIZING MECHANISMS

There are several techniques for converting angular to linear displacements. One of these, used in some low-cost instruments, merely approximates the deflection arc by a straight line (Figure 2-3a). The length of arc subtended by the angle of deflection α is proportional to the angle, while the corresponding semichord is proportional to the sine of the angle. Hence the relative error inherent in this approximation,

$$\frac{\alpha}{\sin \alpha} - 1$$

increases with the angle, becoming 1% at about 14° and 2% at 20°. The reading error in a small recorder is likely to be at least as great as this, so that the linear approximation may be valid.

In recorders using light-sensitive paper or a liquid jet recording mechanism (to be described subsequently), it is common to find the straight line (AB in Figure 2-3b) formed by projection of the angle of deflection α. In this case, the geometry is different; a straight-line deflection recorded along AB is now proportional to $r \tan \alpha$, whereas the true deflection along an arc is proportional to $r\alpha$.

The distortion appearing in curvilinear recordings can be eliminated by a simple lever mechanism which forces the pointer tip to follow a straight line (Figure 2-4). This device, found in

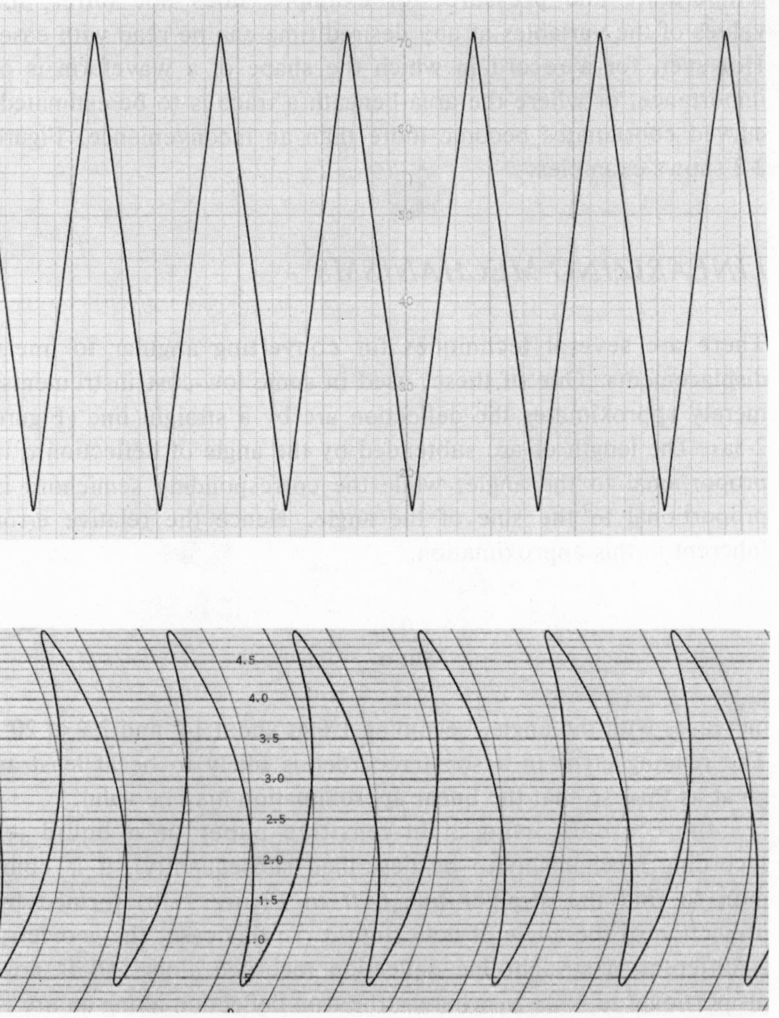

Figure 2-2. A triangle wave reproduced by a recorder whose writing mechanism has been linearized and (below) the same wave as reproduced by the deflection recorder of Figure 2-1(b).

several well-known lines of recorders, permits easy observation of waveforms on a rectilinear grid, but still suffers from the sine error. Both the equal- and unequal-arm devices, as shown in the figure, possess the same error. Choice between these two modifications is one of convenience.

Another principle, used by two prominent manufacturers, is shown in Figure 2-5, and is described in the caption. The lengths of the lever arms and relative sizes of pulley and drum can be selected to reduce the nonlinearity error to less than 0.5% of full-scale.

All recorders discussed thus far are liable to an inherent decrease in sensitivity with increasing deflection because of the nature of the magnetic force producing the deflection. This force is proportional to the current through a coil suspended in a constant uniform magnetic field, and to the cosine of the angle of deflection. Taking into account the cosine error, the rectilinear deflection can be shown to be

$$CI\,\frac{\tan\alpha}{\alpha} \qquad \text{(for Figure 2-3a)}$$

or

$$CI\,\frac{\sin\alpha}{\alpha} \qquad \text{(for Figure 2-3b)}$$

where C is a constant of proportionality, I is the current to be measured, and α is the deflection angle. Ideally, the deflection is independent of α, and for small angles, the error is negligible.

Where greater precision is needed, these errors can be corrected in at least two ways: (1) The magnetic field can be designed to be nonuniform in such a way as to cancel all or part of the error. (2) The recorder may incorporate an input amplifier whose response is nonlinear, i.e., proportional to $\alpha/\tan\alpha$ or $\alpha/\sin\alpha$, as appropriate.

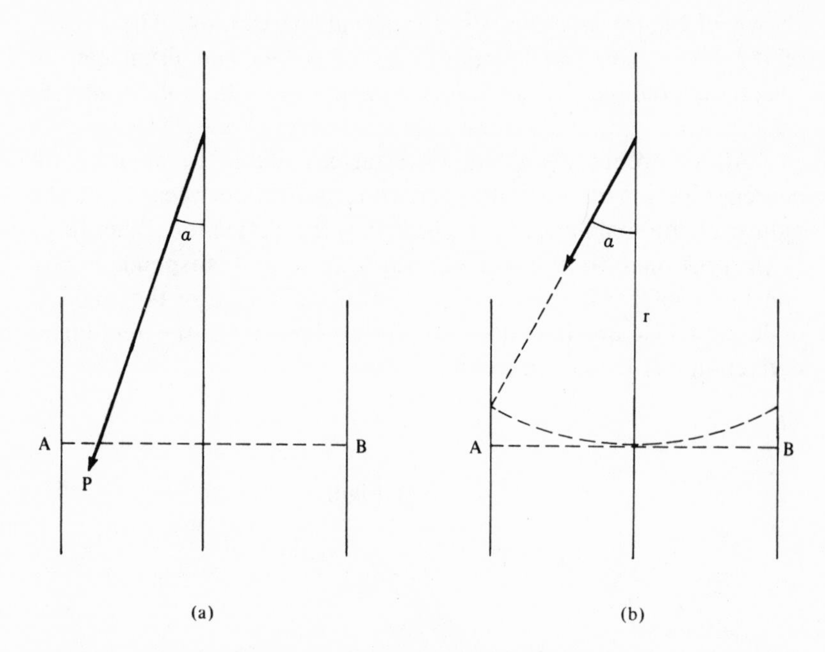

Figure 2-3. (a) The arc described by the tip of the pointer P is approximated by recorded trace $\overline{AB}$. The record is made, e.g., by pressure contact at the intersection. (b) Trace $\overline{AB}$ recorded by projection as in a light beam or liquid jet recording.

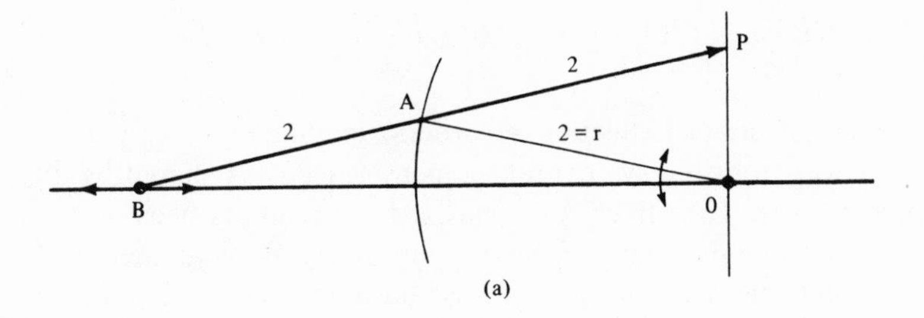

(a)

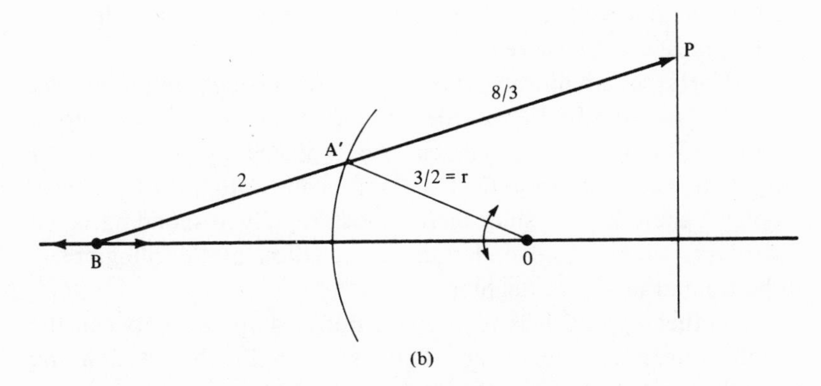

(b)

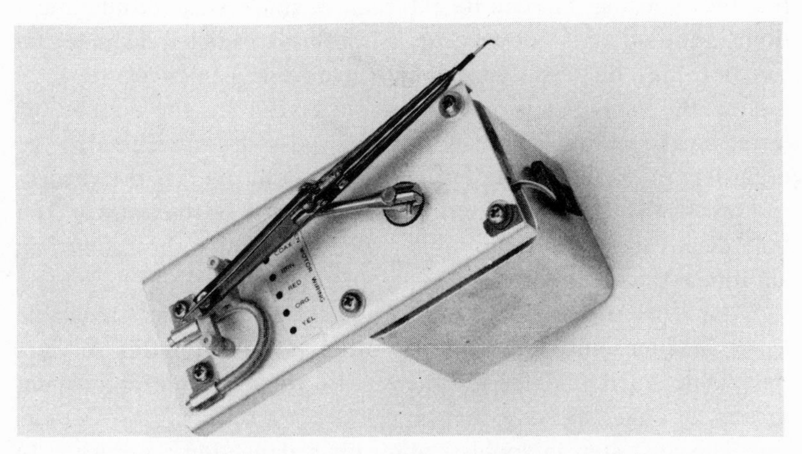

(c)

Figure 2-4. (a) Linearizing mechanism of the equal-arm type and (b) unequal-arm type. Circular motion about point *O* produces straight-line motion of the pointer *P*, perpendicular to the centerline *OB*; the far end of the pointer is constrained to move along the center line at *B*. (c) Linearizing mechanism from a Hewlett-Packard recorder.

THE PREVENTION OF LOADING EFFECTS

The fault of direct deflection recorders, in that they must draw their power from the system under measurement, is shared by all simple meters. The effect is emphasized in recorders because the inertia of moving system, including pen or stylus, is greater, and friction between pen and paper must be overcome. The effect is typically to lower the voltage of the source in the act of measuring that voltage, or to reduce the voltage available to a load, if measuring a current.

Efforts to ameliorate this difficulty through reducing the inertia, mass, and friction in the moving element are fruitful up to a point. The writing system can be replaced by a tiny mirror affixed to the coil, to reflect a light beam onto moving photographic paper. This results in the most rapidly responding of all recorders, widely used under the designation of "oscillograph," to be treated in a later chapter.

Another approach is to insert a buffer amplifier between the signal source and recorder. This is often located within the recorder housing, but can be external. A single stage of amplification, in the voltage-follower configuration, is usually adequate. Its function may be regarded, depending on one's point of view, as raising the impedance of the recorder seen by the source, or lowering the output impedance which the source presents to the recorder, or merely providing more power to operate the recording mechanism without altering the signal to be measured. The only disadvantage (besides some added expense) is the need of an additional power supply for the amplifier.

An "operational" or "instrumentation" amplifier, available as an inexpensive plug-in unit, is often selected for this service. Selectable input and feedback resistors then provide convenient means for changing scale ranges.

The next step in sophistication for a deflection recorder is to provide overall feedback including both the deflection system and amplifier in the feedback loop. This type of servo system is described in the next chapter.

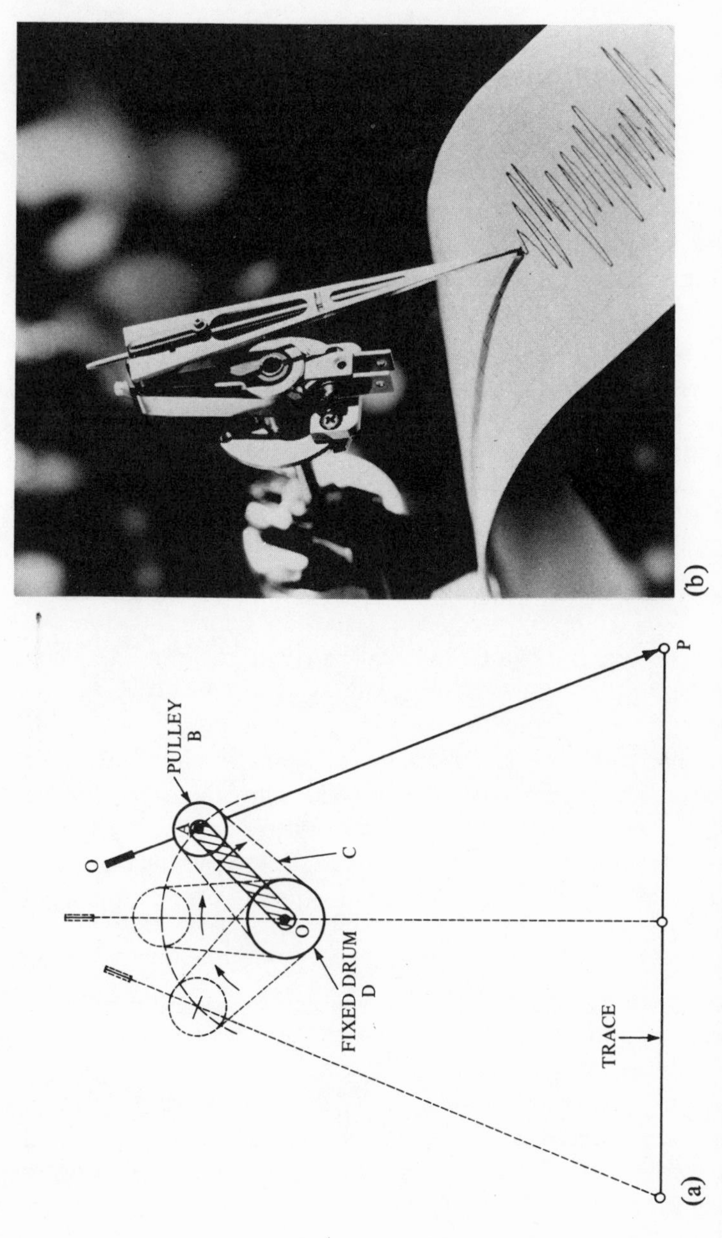

Figure 2-5. (a) Linearizing mechanism used in Gould (formerly Brush-Clevite) and Esterline-Angus recorders. Arm *OA* is attached to the moving coil system and rotates with it, while coaxial drum *D* remains stationary. The outer end of the arm bears a pulley *B* with the pointer rigidly attached to it. A belt *C* passes around both drum and pulley. The relative diameter of drum and pulley, together with the lengths of arm and pointer, are chosen so that the pen *P* describes a straight line across the recording paper for a rotational motion of the penmotor. (b) The writing mechanism from a Gould recorder.

3

Servo Recorders

Probably the most widely used laboratory recorders are those built around a motor-driven servo loop. A motor drives the pen across the chart until a position sensor observes that its location agrees with the requirement of the input signal. There are many possible implementations of this basic principle. The motor, for instance, may be limited to less than one full turn, with direct connection to a pointer carrying the pen or stylus (this is usually thought of as a meter movement rather than a motor), or it may be a more conventional motor, either ac or dc. Those recorders that include two independent servo systems driving the pen parallel to the X and Y axes (X–Y recorders) are discussed in Chapter 4.

A block diagram representing a universal servo recorder is presented in Figure 3-1. The signal voltage (preamplified if necessary) and the reference voltage are both passed through attenuators (to be described below) to a differential amplifier. The output of this amplifier is called the "error signal," since it represents the difference between the true signal and the fraction of the reference voltage corresponding to the pen position. These two voltages should be equal if the recorder is in balance. The error voltage drives the motor in such a direction as to diminish the error, and the motor in turn drives the pen to the point on the chart corresponding to the input signal. The second feedback path, marked "velocity," will be described later.

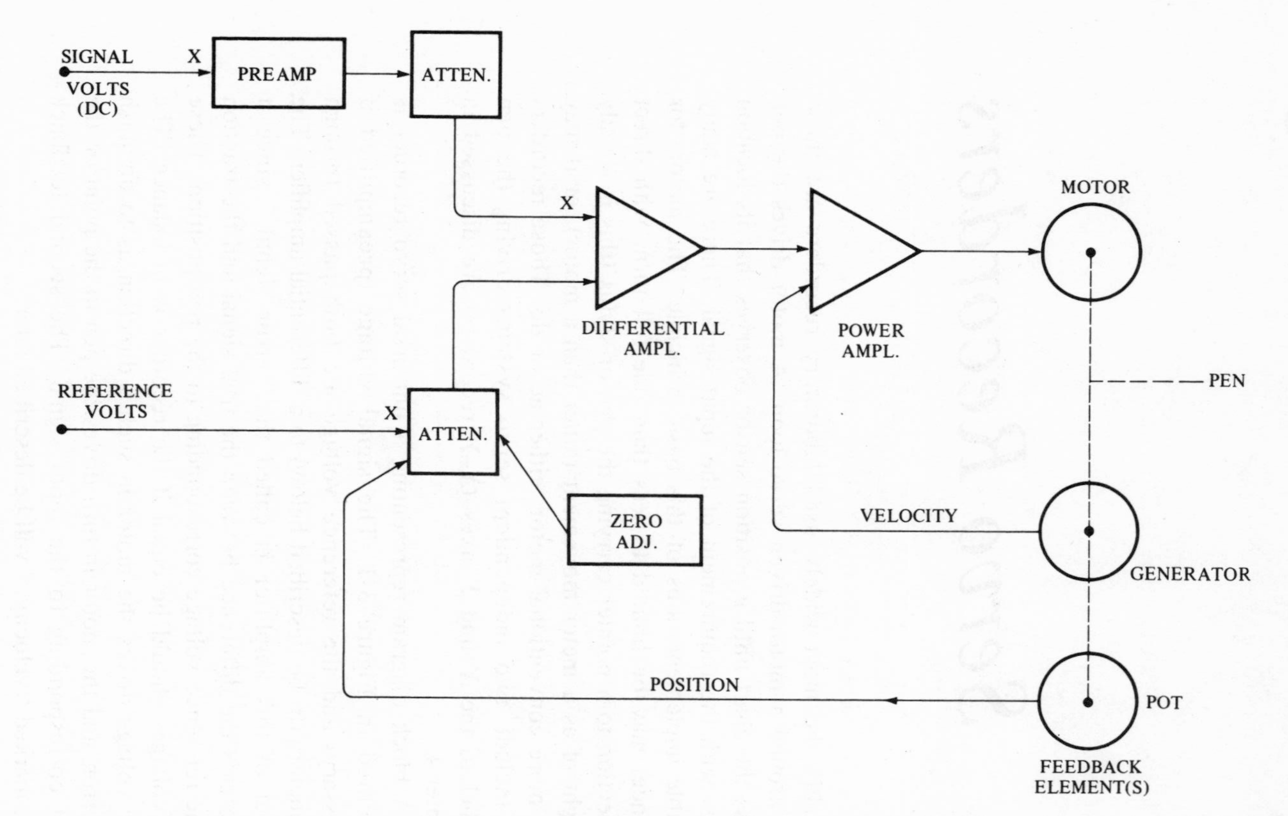

Figure 3-1. "Universal" servo recorder, block diagram. In any particular instrument, some items may be omitted, others combined. The positional feedback can be connected at any of the points marked ×. Details may vary.

Not all the blocks depicted in Figure 3-1 must necessarily appear in all recorders. Many instruments do not include a velocity feedback, and a preamplifier is often not required. The power amplifier may be included with the differential stages in a single unit, the "servo amplifier."

ATTENUATORS

Figure 3-2 shows a typical attenuation circuit. In the most sensitive stages, the input signal is led directly to the amplifier, where it is compared with a measured fraction of the reference voltage picked off from a slidewire. The amplifier must have a high input resistance (impedance). A FET-input unit is suitable, and presents a resistance of the order of 10^{10} ohms or more to the signal.

For higher ranges, the signal must be scaled down by a resistive voltage divider (R_1, R_2). The input impedance is now $R_1 + R_2$, which in our illustration is 1 megohm. The reference voltage must also be divided down so that the slidewire will be supplied with the appropriate potential.

REFERENCE VOLTAGE SOURCES

In older recorders, the reference voltage was supplied by zinc–carbon dry batteries, and frequent calibration was required, i.e., adjustment of the current to validate the stated ranges. A Weston standard cell was usually incorporated for this purpose. In some more recent recorders, mercury cells have been substituted; the mercury cell maintains its stated voltage throughout its life, and calibration is only necessary much less frequently. In present-day designs, all batteries are eliminated, and constancy of reference voltage is assured through the use of temperature-compensated zener diodes. Calibration is seldom required after initial factory adjustment.

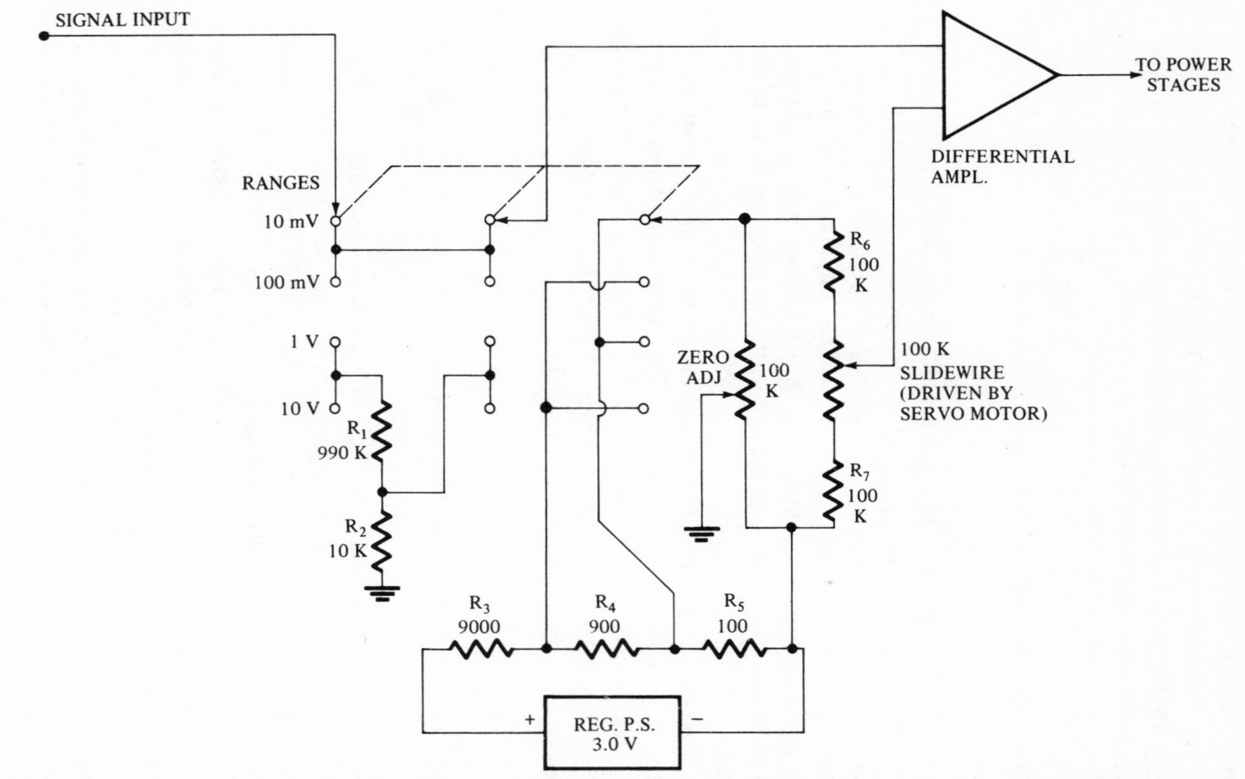

Figure 3-2. A typical input attenuator circuit for four ranges (10 mV, 100 mV, 1 V, and 10 V, full-scale). A three-segment switch with four contacts in each segment is required.

SLIDEWIRES

The slidewire, in servo parlance called a "rebalancing pot," takes many forms. In older instruments, and some recent designs as well, it consists of a single turn of resistance wire wound on an insulating drum 4–6 inches in diameter (Figure 3-3). Alternatively, the wire can be wound into a long, tight helix, and the helix mounted on the drum. The wiper in these designs consists of a spring-loaded metal contactor on an arm rotatable around the center of the drum. The contact should be made of a metal softer than the slidewire, since contactors are cheaper to replace than precision slidewires. The whole assembly must be protected from dust and other contamination.

Since precision multiturn potentiometers have become commercially available, many recorder manufacturers have turned to them as an economical and convenient slidewire. These are mostly compound helices mounted on the inside of a hollow plastic cylinder, provided with a spring wiper, and hermetically sealed.

Another type of slidewire takes a straight-line geometry rather than a circular winding. It is conveniently mounted parallel and adjacent to the track bearing the pen, so that a single carriage can carry both pen and wiper. This straight resistive element can be a simple wire or a tight helix, or it can be a strip of resistive plastic material. The latter has become quite popular because it is highly resistant to wear, and is easily cleaned.

The resolution possible with different types of slidewires varies considerably. Any design in which the slider makes contact with successive turns of a helix is only capable of giving a stepwise approximation to the desired potential variation. These steps are usually well below 1% of full scale, and may not be noticeable. A single-strand wire does not show this effect, but in order to have enough resistance, may have to be very thin, hence especially liable to damage. Plastic strips also show infinite resolution.

Two sources of trouble with the slidewire can be greatly reduced through the use of a small operational amplifier as a

Figure 3-3. The slidewire from a Leeds and Northrup potentiometric recorder.

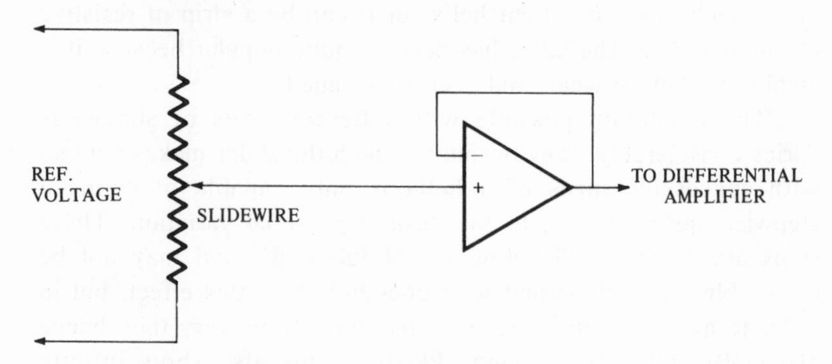

Figure 3-4. An operational amplifier to eliminate the possibility of drawing current through the sliding contact of the rebalance slidewire. An amplifier such as μA709 (Fairchild) or LM101 (National Semiconductor) is suitable.

voltage follower, as in Figure 3-4. The purpose of this amplifier is to prevent appreciable current from flowing through the sliding contact of the potentiometer. This improves the linearity by eliminating loading effects, and also markedly decreases noise due to varying contact resistance, since such resistance is negligible compared to the input resistance of the operational amplifier.

A zero-adjustment potentiometer is conveniently connected in parallel to the slidewire, as in Figure 3-2. This will determine the pen position when the signal input is shorted to ground. It can be adjusted to any desired point, for instance, at either end of the scale or in the center. The two resistors R_6 R_7 in series with the slidewire make it possible to set the zero point as much as the equivalent of one full-scale span either above or below the scale. The scale displayed on the recorder (and printed on the paper) customarily has its zero point at the left. For laboratory use, however, it may be convenient to have the zero at the right, so that the chart will assume the usual graphical format when removed from the recorder, namely the origin at the lower left, time increasing to the right, and the displayed function vertically upward.

It is possible to interchange the roles of signal and reference voltages, impressing the signal across the slidewire and equating a measured fraction of it with the reference. In this arrangement, a preamplifier is essential to ensure that the slidewire voltage will be larger than the reference, and to keep current from being drawn from the signal source.

AMPLIFIERS

The older vacuum-tube amplifiers have been completely replaced in new designs with solid-state amplifiers, either using discrete transistors or integrated circuits. Separate output transistors are usually required to provide enough power to control the motor. In many recent recorders, the electronic circuitry is mounted on plug-in printed-circuit boards. Some manufacturers have chosen to use plug-in amplifiers made by one of the many component

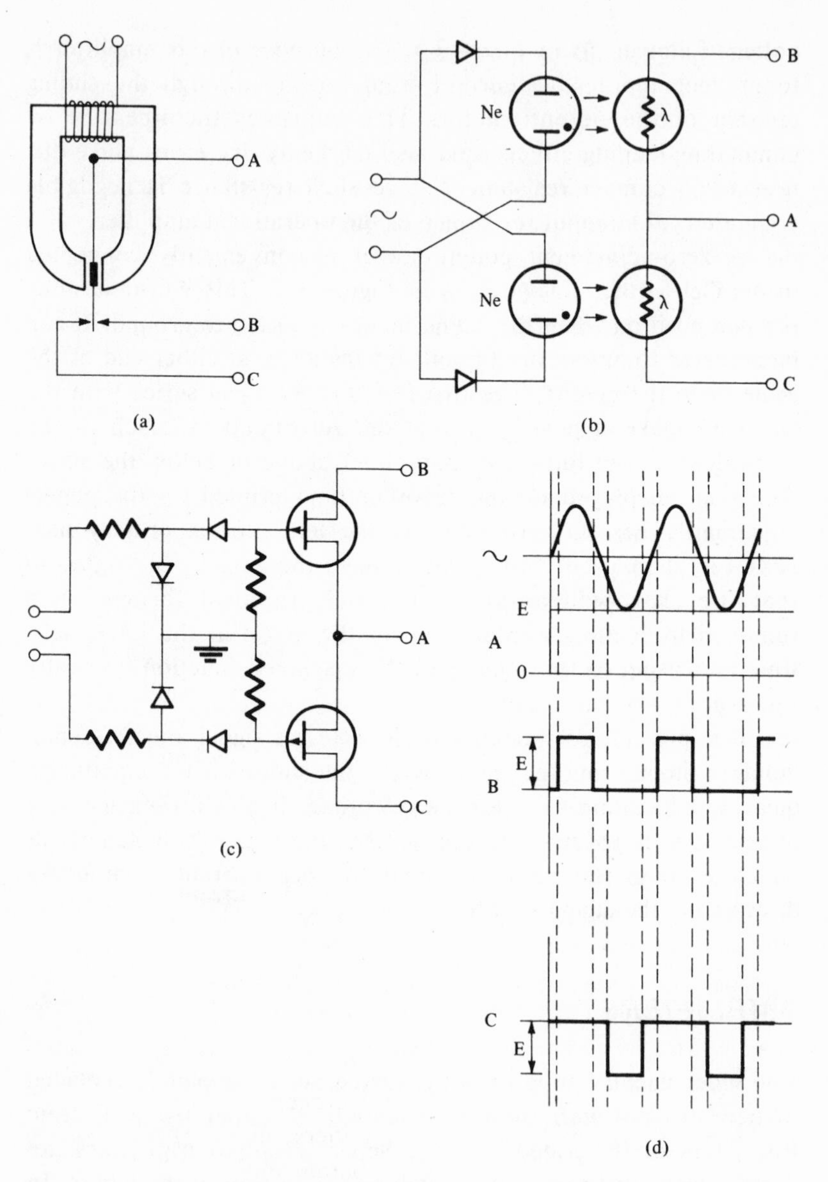

Figure 3-5. Chopper (modulator) circuits. (a) Mechanical vibrator; (b) photo-chopper utilizing neon lamps and photo-conductive cells; (c) transistor chopper with FET switches; (d) waveforms at the several terminals, assuming a dc potential of E volts at the input A.

companies specializing in electronic subassemblies. The principal requirement is a high-impedance differential input. For most servo recorders, the speed of response of standard solid-state amplifiers is much greater than that of the moving system, and so this factor need not be considered.

In the recorder diagrammed in Figure 3-1, it is assumed that the amplifier responds directly to dc potentials. In many instruments, however, the dc signal is converted to ac before being presented to the amplifier. This is accomplished by means of a mechanical chopper or its equivalent. Figure 3-5 shows some examples, in the SPDT configuration. (There are many alternative circuit arrangements to accomplish the same objective.) As shown in Figure 3-5(d), a dc potential of E volts is converted into an approximation of a square wave with amplitude E. (The output is generally introduced into the amplifier through a transformer, and this rounds off the corners of the square wave, so the secondary of the transformer produces a fair approximation to a sine wave.) Figure 3-5(a,b,c) shows different ways of chopping the signal to produce the same waveforms.

Figure 3-6 indicates how such a chopper can be used in a servo system. The impedance seen by the signal is no longer the input resistance of the amplifier proper; rather, it is a variable quantity depending on the nearness to balance. If the signal and reference voltages are exactly equal, then no current can flow, and the resistance effectively approaches infinity. When the system is "slewing," i.e., seeking a new balance, the input impedance is less in proportion to the departure from balance.

There are several advantages to converting the signal to ac for amplification. One is that slow zero-point drift, which plagues dc amplifiers, is eliminated by the inductive or capacitive coupling associated with ac amplifiers. Another advantage is the ease with which the output from the amplifier, if chopping is at line frequency, can be used directly to control the response of a two-phase ac servo motor.

Chopping at line frequency is not without drawbacks, however. Very careful shielding and high-quality components are mandatory to avoid pickup of line-derived noise, whereas if the

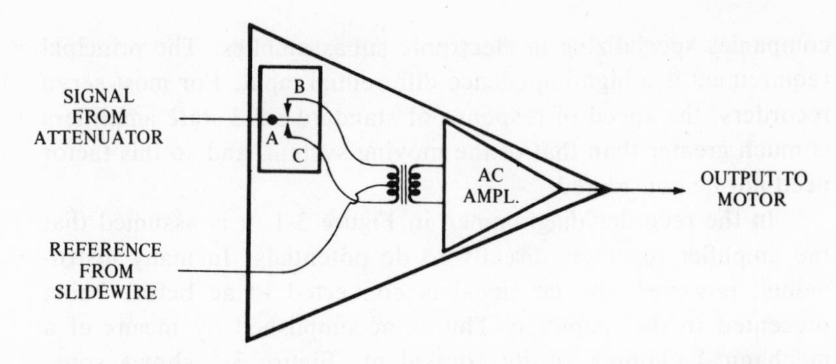

Figure 3-6. **The use of a chopper in the input circuitry of a servo recorder. The letters *A, B,* and *C* correspond to those in Figure 3-5; the large triangle is the same as the small one in Figure 3-2.**

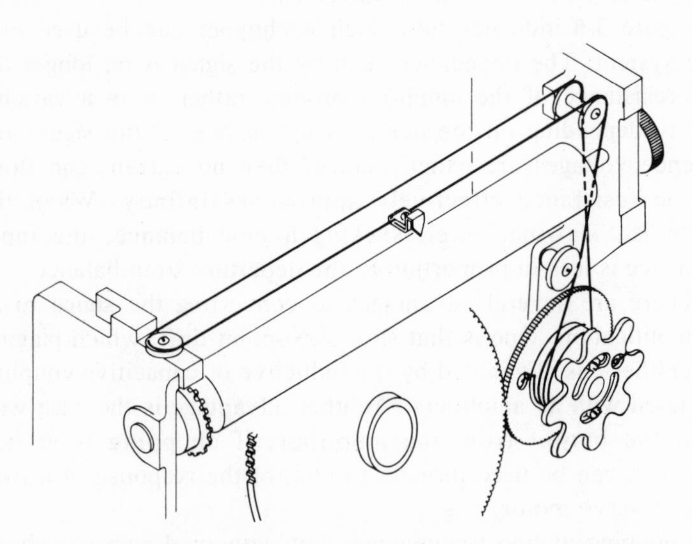

Figure 3-7. **A typical stringing diagram. The handle provides a means of adjusting the tension in the string.**

signal were chopped at some other frequency, a simple *RC* filter could be used to reject 60-Hz noise. The frequency response of the recorder is limited by the chopping frequency. The use of a local oscillator to produce a higher chopping rate is difficult if a two-phase motor is employed, because the oscillator must provide sufficient power to energize the reference winding. A solution to this dilemma, adopted by at least one manufacturer, is to dispense with the two-phase motor in favor of a dc motor. A phase-sensitive rectifier will provide a dc current in one direction if the error signal is positive, in the other if negative, so the motor will turn in the appropriate directions.

MOTORS

As previously mentioned, the servo motor may be either dc or two-phase ac, the chief requirements being reversibility and low inertia, so that it can follow with precision any commands that it receives. Most servo recorders use conventional rotary motors connected to the slidewire and pen through mechanical linkages. This necessitates the use of a cord or cable with the pen carriage firmly fixed to it, wound around a driving drum. (An alternative lead-screw drive would be too slow for most recorders.) The layout of the cable drive requires careful design if it is to be trouble-free and easily replaced (Figure 3-7). Stringing a new cable can be a most frustrating task for a nonspecialist!

To avoid such difficulties, and in the interests of general simplification of design, a number of manufacturers have developed motors in which the prime motion is linear rather than rotational. Figure 3-8 shows an example.

In connection with the moving servo mechanism, mention should be made of certain related mechanical features. The moving system must be protected from overdrive. This is sometimes accomplished in the electronics through amplifier limiting, but can be achieved mechanically, either by limit

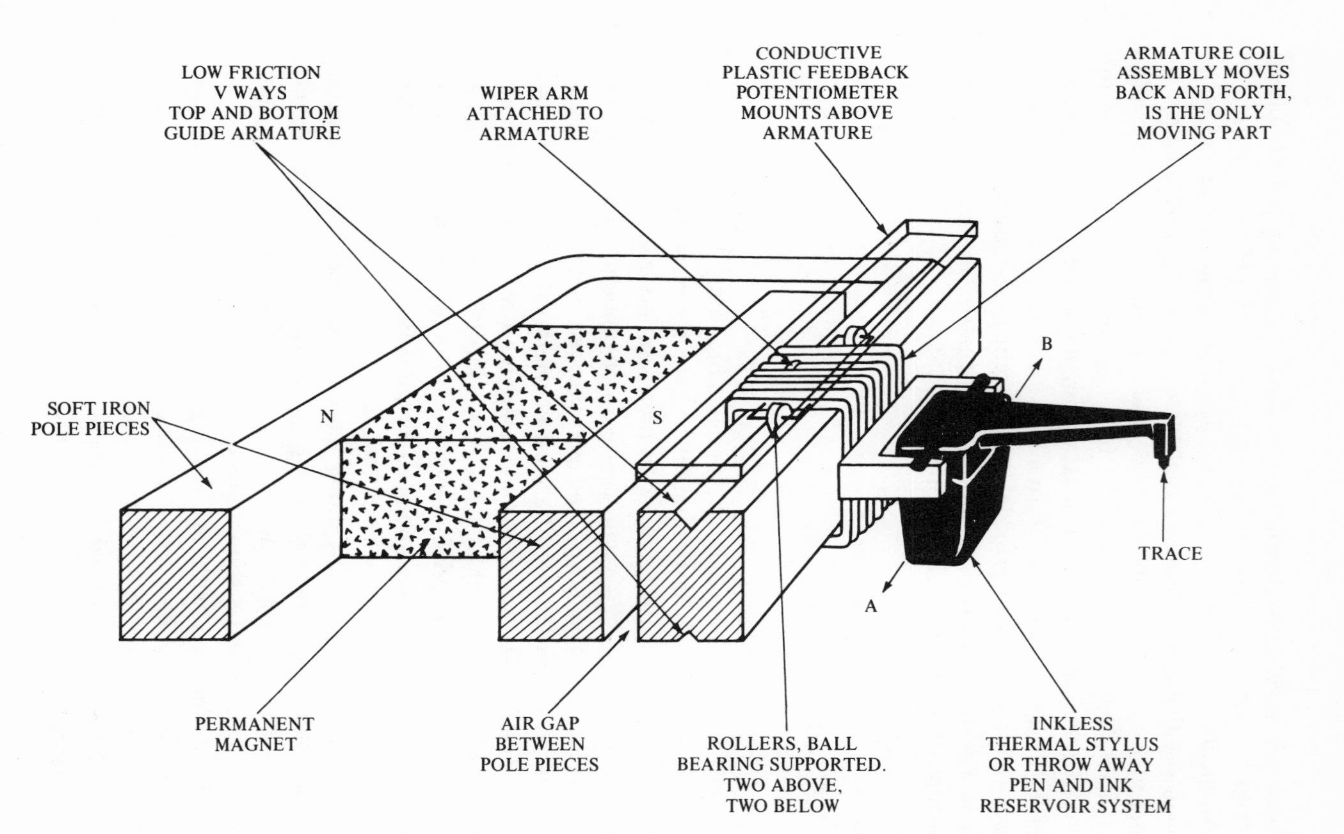

Figure 3-8. Linear servo motor of the permanent-magnet dc type (Esterline-Angus).

switches or a slip clutch. In the first alternative, the carriage closes a miniature switch at each end of its safe travel, thus cutting off motor power. The disadvantage is that power must be restored somehow when the signal drops to safe levels, and this is not simple to do. Hence the friction clutch is more favored even though it is prone to wear.

Another mechanical feature that occasionally causes trouble is the method of making electrical contacts with moving parts. If the moving element is constrained to less than a single turn (e.g., a meter movement) or its linear equivalent, connection is often made by slack cables or pigtails of flexible wire. If many turns are required, then a system of slip-rings similar to those in a conventional ac generator is necessary.

MULTICHANNEL RECORDERS

For many applications, it is desirable to record two or more functions simultaneously on the same time scale. Of course, this can be accomplished with multiple synchronized recorders, but more economically with a single recorder provided with multiple independent servo systems. The several channels share the same paper drive mechanism and power supplies, but must be provided with separate range and sensitivity controls. A choice must be made between overlapping operation and side-by-side location, where each channel uses only its share of the useful width of the paper. In the latter, the reading error is multiplied by the number of channels, while in the former, the time base must be slightly offset between channels so that the pens can pass each other. Figure 3-9 shows examples of the two alternatives.

The side-by-side type requires no major changes in design, but overlapping pens present some new features, which have been solved by various ingenious means. The trick is to be able to operate each pen carriage without interference, and yet to keep the time offset minimized. Versions of both types are offered by most major manufacturers. One company (Esterline-Angus) has a

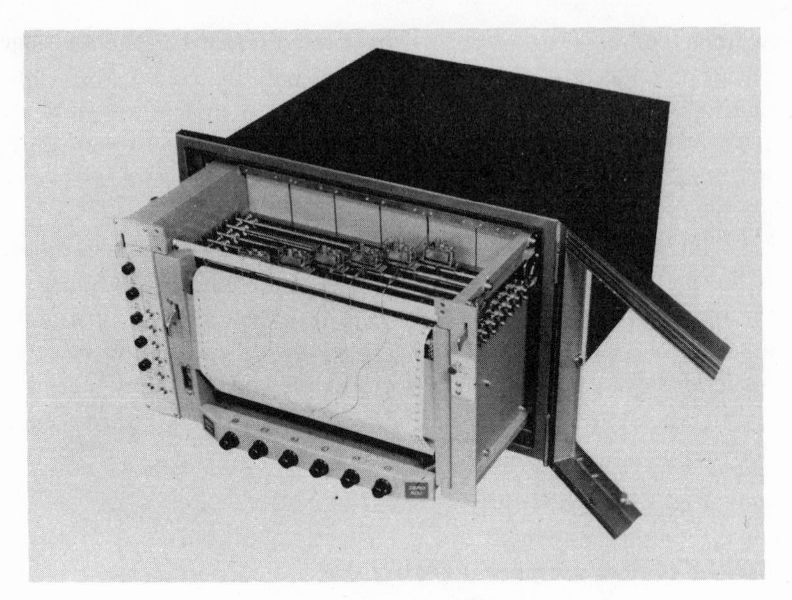

(a)

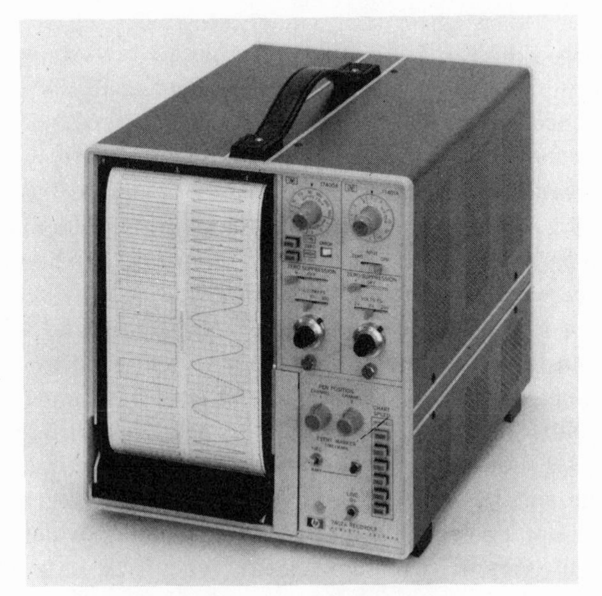

(b)

Figure 3-9. (a) Overlapping multichannel recorder (Soltec Co.). (b) Side-by-side operation (Hewlett-Packard).

two-pen overlapping model of their linear motor recorder; the design problem of mounting two linear motors running parallel to each other is considerable.

DYNAMIC PROPERTIES

The simple servo system with position-sensing feedback may not always show the response desired, especially if called on to adjust rapidly to changes in input. The three types of response likely to be observed are shown in Figure 3-10. The underdamped response (a) appears as an oscillation of diminishing amplitude, known as hunting or ringing. At the other extreme is curve (b), corresponding to overdamping, a sluggish response. Curve (c) displays critical damping, the least which will prevent overshooting. Usually the optimum is a slight underdamping, as in curve (d), where the slight overshoot is acceptable, and the response has reached the zero-error value somewhat sooner than in (c).

Damping can be controlled mechanically with a dashpot, electrically with an *RC* network of high time constant, magnetically by the frictionless drag produced by eddy currents, or dynamically by means of the velocity feedback loop indicated in Figure 3-1. In the latter, some device must be incorporated in the moving system which will produce an electrical signal proportional to the velocity. This is often a small ac tachometer generator with line frequency and with a 180° phase shift between rotation in opposite directions. This ac voltage can be applied to the reference connection to the differential-input power amplifier stages. (The phase of the ac applied to the excitation winding of the tachometer may need adjusting to make this system work.) The power delivered to the servo motor will now be diminished from what it otherwise would be at high speeds but not changed appreciably at low speeds near balance, thus increasing the dynamic stability. The amount of feedback is easily adjusted by a variable resistance.

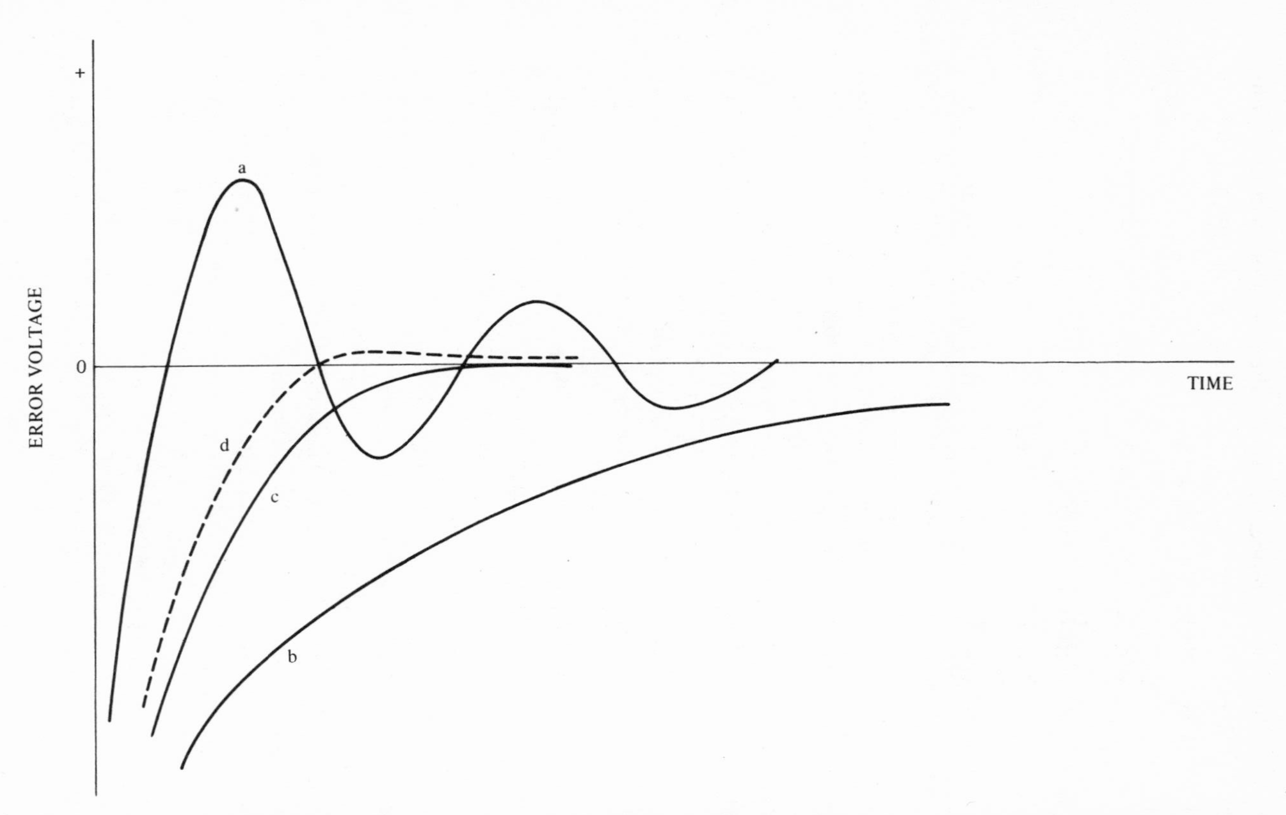

Figure 3-10. Dynamic response of a servo system to a step perturbation: (a) underdamped, (b) overdamped, (c) critically damped, (d) optimum, slightly underdamped.

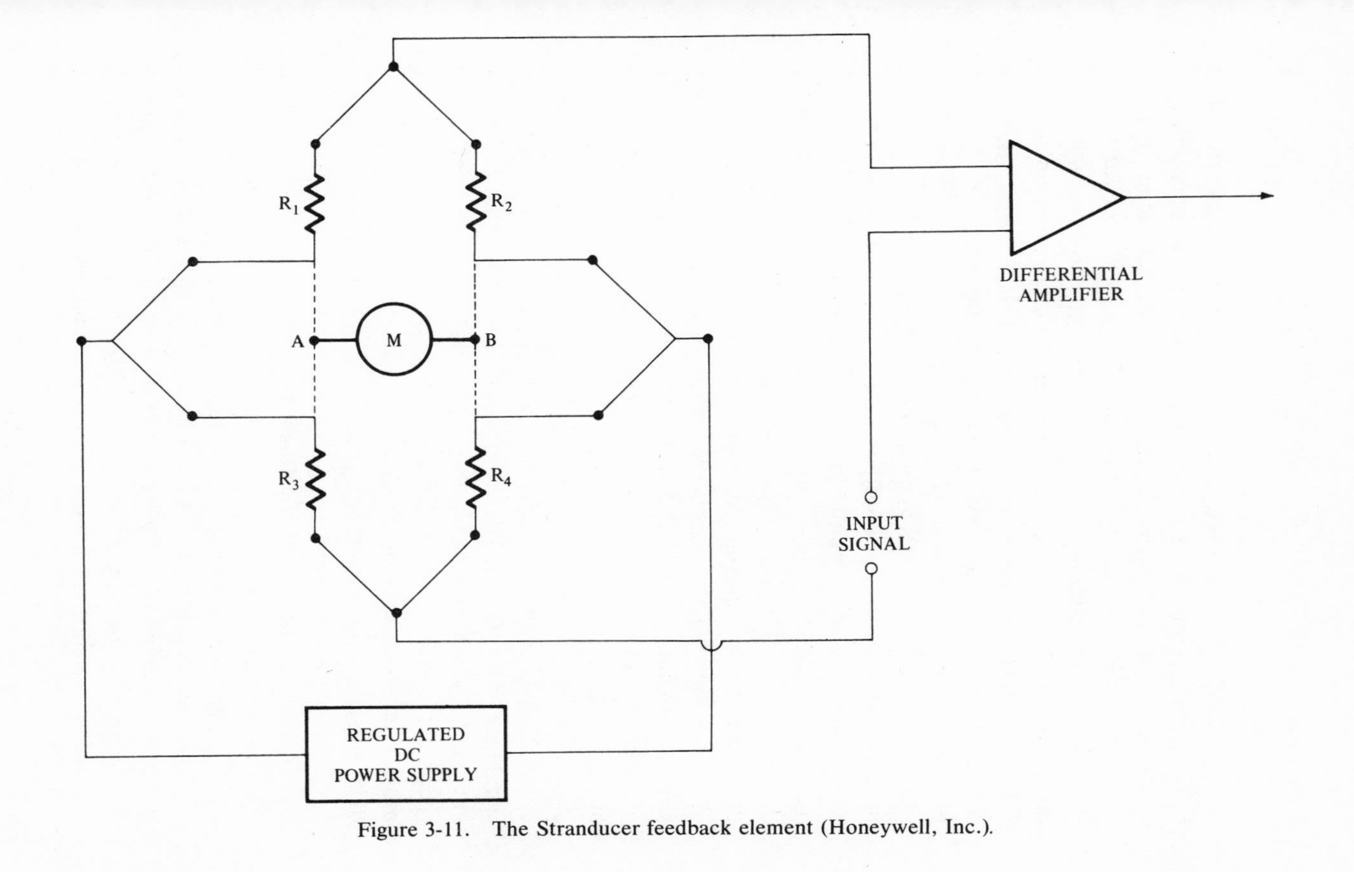

Figure 3-11. The Stranducer feedback element (Honeywell, Inc.).

SPECIAL MODIFICATIONS

As might be surmised, there are many modifications of the basic servo system. Some of these, of course, are proprietary. A few with features of particular interest will be described here.

Honeywell, in their "Electronik-17," have dispensed with the moving contact slidewire as a feedback element, substituting a device which they call a "Stranducer," based upon an array of four unbonded strain gages (Figure 3-11). In this figure, M represents a shaft geared down from the servo motor, and provided with a cross-arm AB. The ends of the arm are made fast physically (broken lines) to four identical strain gages connected in a Wheatstone bridge configuration. If the motor tends to turn counterclockwise, tension is increased in R_1 and R_4, which increases their resistance, and diminished in R_2 and R_3, which decrease in resistance. The unbalance voltage at the bridge diagonals is thus determined in magnitude and sign by the rotation of the motor, and this feedback voltage is continuously compared with the signal, as usual. This device is said to have a far longer life expectancy than the conventional slidewire. Its accuracy is about the same.

Several companies offer servo recorders wired as self-balancing Wheatstone bridges. Figure 3-12 is an example. The rebalance pot, instead of acting as a reference voltage divider, serves to keep the bridge continuously in balance. Bridge recorders can be used with any resistive transducers, but their most extensive use appears to be recording temperature by means of a resistance thermometer.

For some purposes, for example, chemical spectrophotometry, it is desirable to plot the logarithm of a function against time. This can be done with a logarithmic amplifier, or with specially cut logarithmic gears (Figure 3-13).

The servo principle can also be applied to recorders in which the pen is moved directly by a coil of wire turning in a magnetic field, the so-called penmotor, the modified meter movement described in Chapter 2. Figure 3-14 shows an example of such an

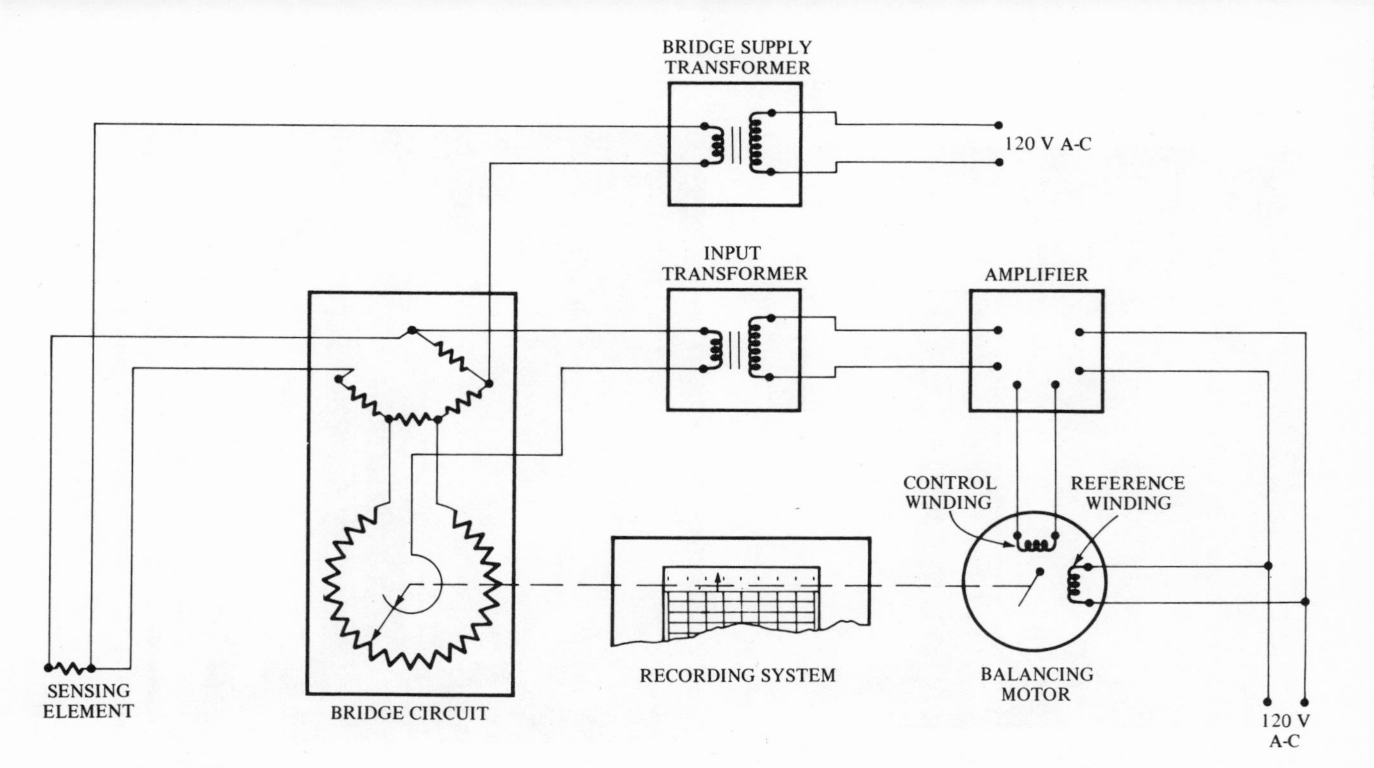

Figure 3-12. **The Dynamaster ac bridge (Bristol). The balancing motor is coupled to the pen and a potentiometer which keeps the bridge circuit in balance.**

(a)

(b)

Figure 3-13. (a) Ordinary gears which cause a linear response of the pen; (b) logarithmic gears (Sargent-Welch).

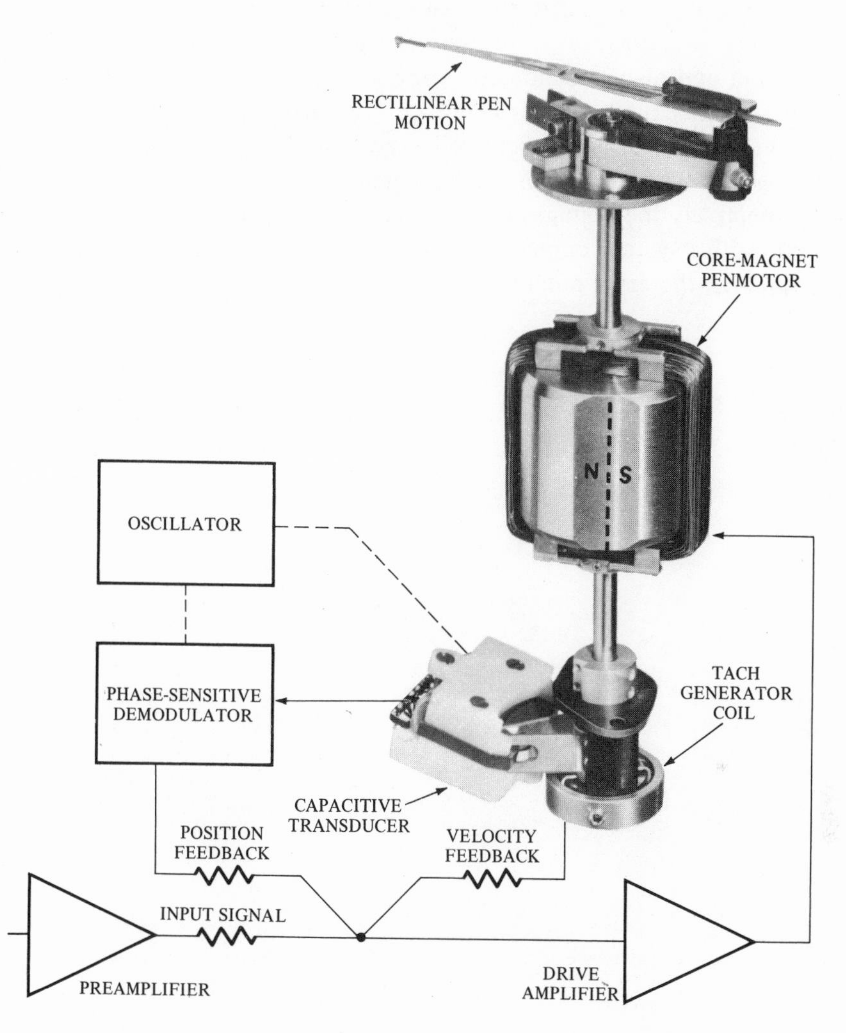

Figure 3-14. Servo control system as used in the Gould Mark 260 Oscillographs. The penmotor coil moves in the radial field of the permanent-magnet core. The position sensor is a capacitive transducer which delivers alternating current with its phase determined by the direction of deflection and its amplitude proportional to its magnitude. A phase-sensitive demodulator converts this ac to dc of sign corresponding to phase, and voltage corresponding to deflection. The linearizing mechanism is of the belt and pulley type described in Figure 2-5.

instrument. Attached to the shaft of the penmotor is a device for sensing the position or extent of angular displacement of the system and also a tachometer generator. The dc outputs of these are summed with the signal to be measured at the input to the drive amplifier. The current delivered to the coil is proportional to $E_s - (E_v + E_p)$, where E_s is the signal voltage, and E_v and E_p are respectively the voltages from the tachometer and position sensor. When the system comes to rest, E_v becomes zero, and E_p must equal E_s; the error potential is zero.

4

X–Y Recorders

By the term *X–Y* recorder is meant an instrument that is capable of plotting any two variables against each other, rather than against time. This requires that the recorder be able to plot multiple-valued as well as single-valued curves. Nearly all *X–Y* recorders are potentiometric servo types, with separate input networks and amplifiers for the two axes. Hence the chief differences from the instruments discussed in Chapter 3 are in mechanical rather than electrical features.

DRIVE MECHANISMS

There are many possible arrangements that can be made to effect the desired relative movement of pen and paper. Those that are currently in use most commonly require individual sheets of paper held on a flat bed. The paper is traversed by an arm moving parallel to the *X* axis while the pen travels along the arm in the *Y* direction. (See Figure 4-1.)

Other possibilities include (a) two moving arms parallel to the two axes, with the pen carriage sitting on their intersection (Figure 4-2); (b) paper held on a concave cylindrical surface; the pen is mounted radially on an axial arm which turns for the *Y* motion, while the pen moves along it for the *X* (early Moseley recorders); (c) a modification of a standard strip-chart recorder in

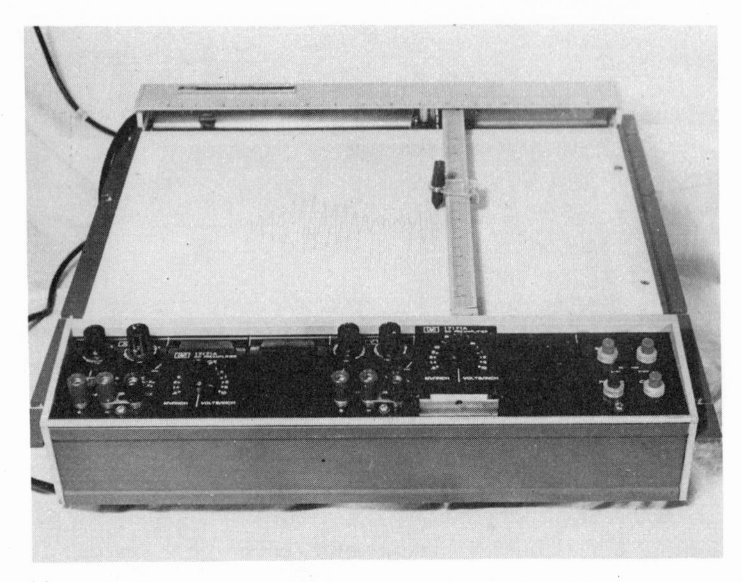

(a)

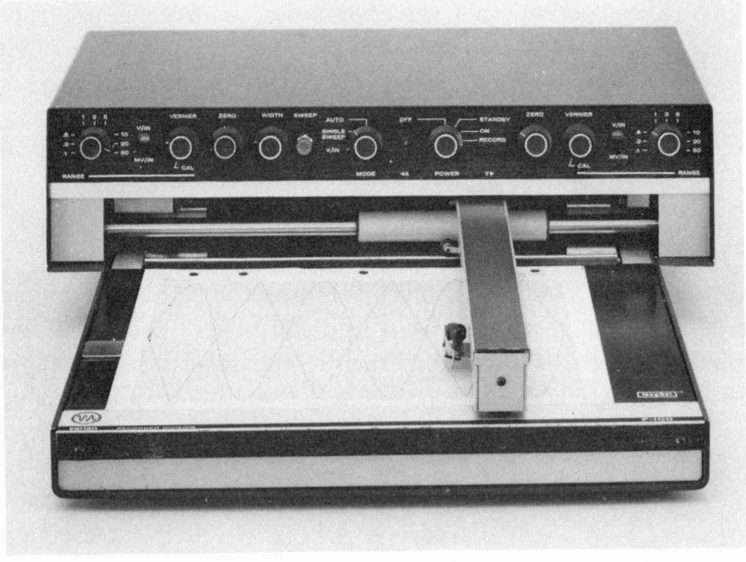

(b)

Figure 4-1. *X*-*Y* recorders. Pen moves along *Y*-axis arm, while the arm is displaced along the *X* axis. (a) Hewlett-Packard Inc. (b) Varian Associates.

Figure 4-2. The pen carriage rides on the intersection of the two moving arms parallel to the X and Y axes. (Houston Instruments Co.)

which the paper can be driven either forward or backward beneath the pen by a second servo system replacing the constant-speed paper drive.

There are a number of mechanisms by which an arm can be made to move across a plane. Two of these are shown in Figure 4-3. It is more complicated to drive the pen along this movable arm; a few mechanisms for this function are depicted in Figure 4-4. The products of several companies utilize the arrangement in Figure 4-4(a), where the *Y* motor is mounted on the cross-arm and hence moves with it in the *X* direction. This has the advantage of eliminating the complex linkages needed if the motor is to be stationary, but increases the inertia of the system driven by the *X* motor. It also multiplies the number of flexible electrical connections that must be provided.

INPUT CIRCUITRY

Many recent *X–Y* recorders are provided with modular plug-in attenuators and amplifiers. The Hewlett-Packard Model 7004 is typical (Figure 4-1). The following modules can be used in either input:

1. DC preamplifier, with 14 calibrated ranges from 0.5 mV to 25 V/in.
2. DC attenuator, with eight calibrated ranges from 0.1 to 20 V/in.
3. Time base, to permit making *X–T* or *Y–T* recordings; has eight speeds from 0.5 to 100 sec/in.
4. Null detector; this converts the recorder to a point plotter. For every specified *X* setting, it waits until servo balance is reached before recording a dot.
5. Scanner; this unit provides rapid switching between two input voltages on the same axis, printing a close-spaced series of dots for each. This is analogous to dual trace operation of an oscilloscope. Both this and the preceding module require replacing the pen with a special point plotter.

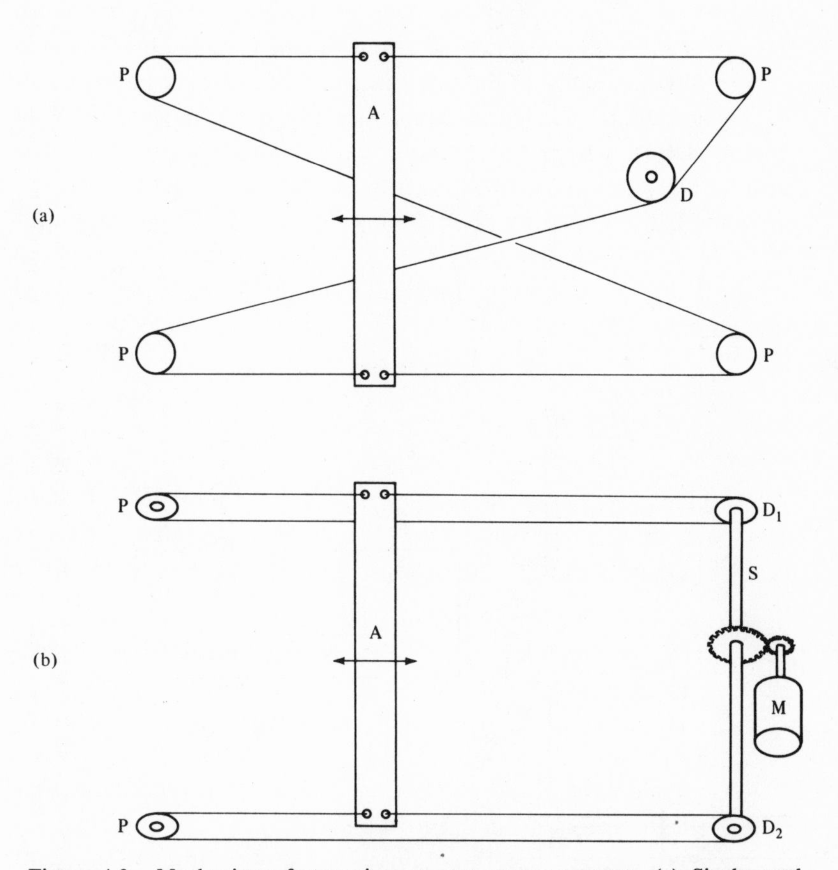

Figure 4-3. Mechanisms for moving an arm across a paper. (a) Single cord, clamped to the arm at both its ends. (b) Two cords. P = pulley, D = driven drum, M = motor, A = arm.

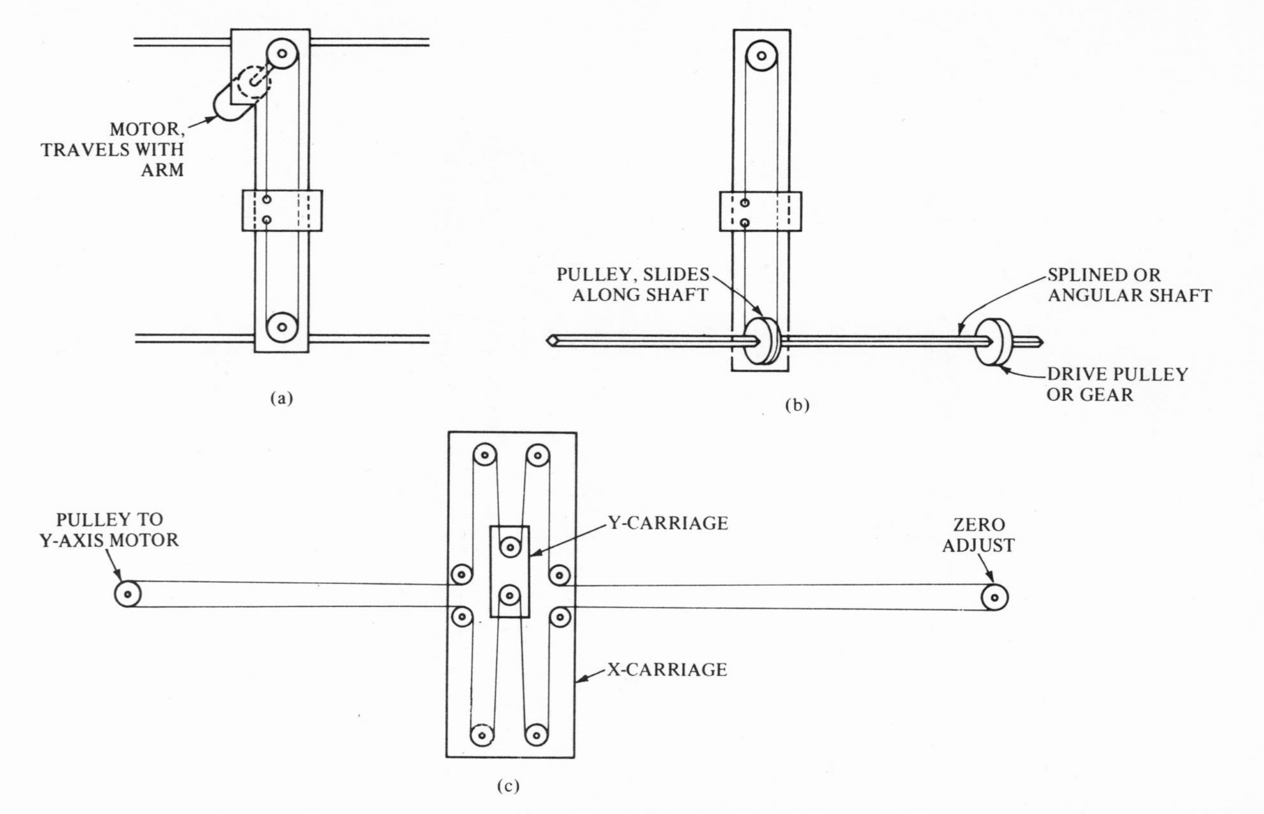

Figure 4-4. Mechanisms for positioning a pen carriage on a movable arm. (a) Motor moves with the arm. (b) Motion transmitted through a sliding pulley on a splined or angular shaft. (c) Multiple-pulley arrangement. In all, the arm or *X* carriage can move right to left without affecting the pen position on the *Y* axis.

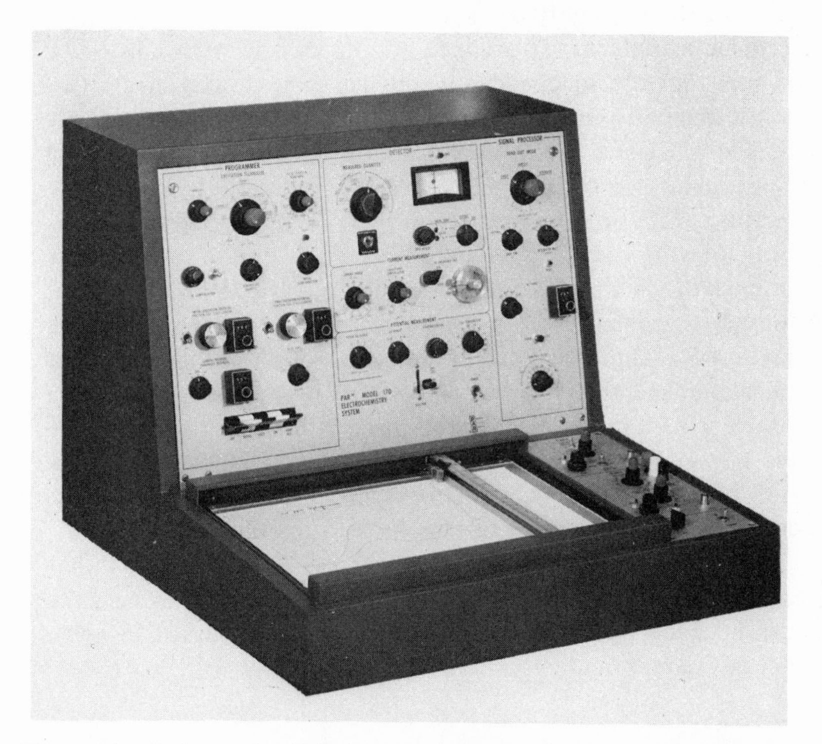

Figure 4-5. *X–Y* recorder built into an electrochemical instrument (Princeton Applied Research Corp.).

6. AC-rejection (low-pass) filter.
7. DC offset voltage, variable from 1 mV to 1 V, either polarity.

Recorders lacking the modular construction often can be ordered with various options, including many of the above features and others. The built-in time base is especially useful in increasing the versatility of the recorder.

Some manufacturers produce "X–Y_1–Y_2" recorders, in which two pens operate independently on the same X-axis arm structure. It is almost essential that the two motors and the corresponding rebalance pots ride on the arm, to avoid interference of multiple pulleys and cords.

Many laboratory instruments of such varied nature as spectrophotometers, polarographs, and tensile testers are equipped by their manufacturers with integral X–Y recorders, often obtained as O.E.M.* items from recorder companies (Figure 4-5). These will have been optimized for the particular application, and their controls combined with those of the main instrument as appropriate. The X drive is commonly run by the same motor that controls the independent variable of the instrument. Many recorder manufacturers offer special stripped-down models with a large variety of options for such service.

A special feature required by those X–Y recorders that accept individual sheets of chart paper is some device to hold the paper in place without slipping. This will be discussed in some detail in Chapter 6.

* Original Equipment Manufacturers.

5

Oscillographs

Analog recorders designed to accept signals that change more rapidly than a servo system can handle are known as oscillographs. The name presumably has derived from analogy with the cathode-ray oscilloscope, which is also suited for high-speed signals; indeed, much higher frequency signals can be displayed on an oscilloscope than on any recorder with moving parts. Most oscillographs fall into one of two classes: those that use photographic recording, and those that rely on pen-and-ink, electrical, or thermal writing methods. Figure 5-1 shows a representative style of multichannel oscillograph.

LIGHT-BEAM OSCILLOGRAPHS

Recorders of this class utilize a sensitive galvanometer provided with a small mirror. As the galvanometer coil turns in a permanent magnetic field, the mirror deflects a beam of light across the width of a moving strip of photosensitive paper (Figure 5-2). After suitable photographic development, the paper shows a trace representing the time-variation of the signal.

Light-beam recorders are often made in multiple-channel models, with a separate galvanometer for each channel. The galvanometer movements are replaceable and interchangeable, fitting into slots between the pole pieces of a single magnetic

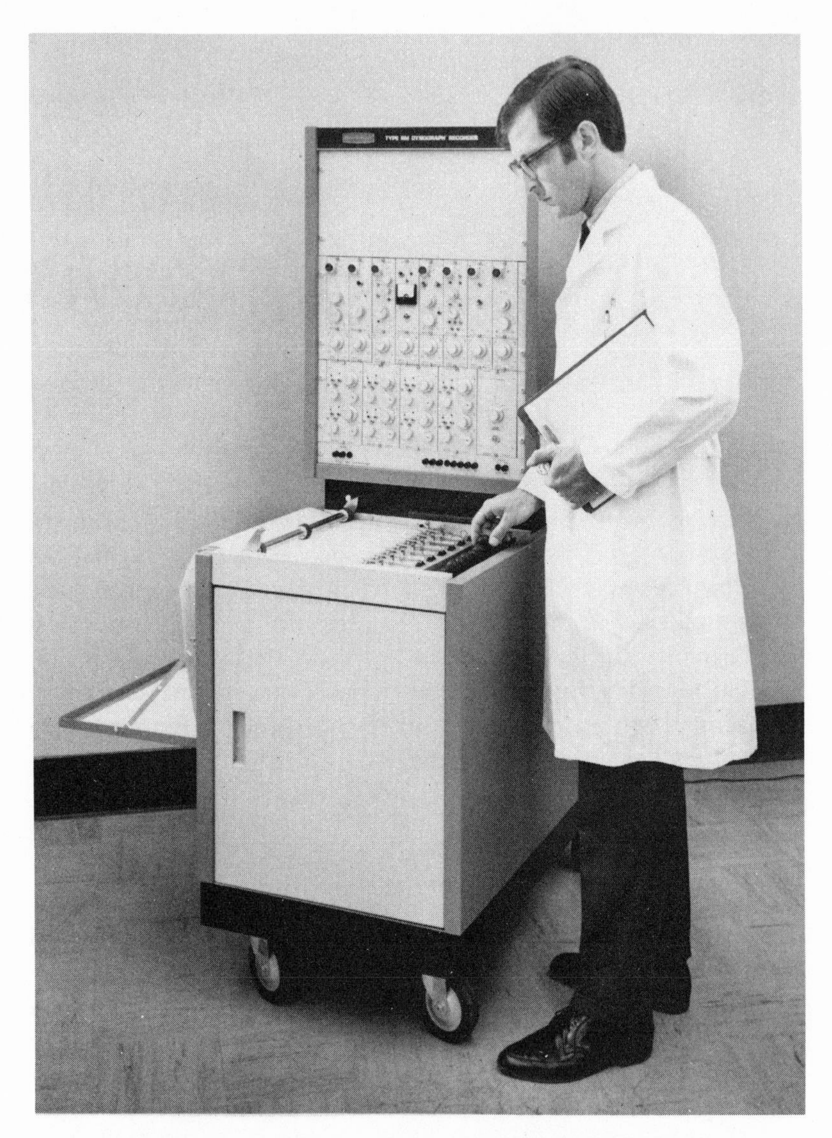

Figure 5-1. Beckman Instruments oscillographic recorder.

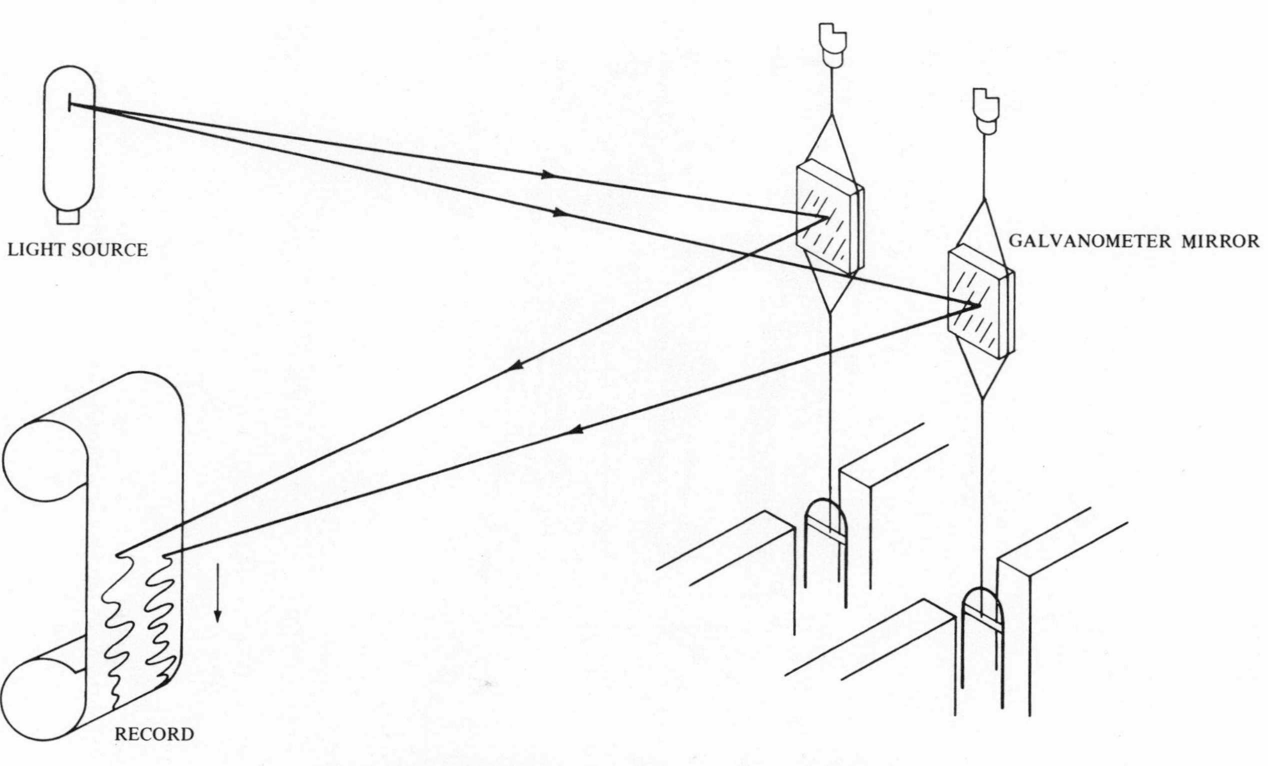

Figure 5-2. Light-beam oscillograph with two channels shown.

(a)

Figure 5-3. (a) CEC optical oscillograph; (b) diagram of the optical system of the CEC Model 5-124 (now manufactured by DuPont). See the text for explanation.

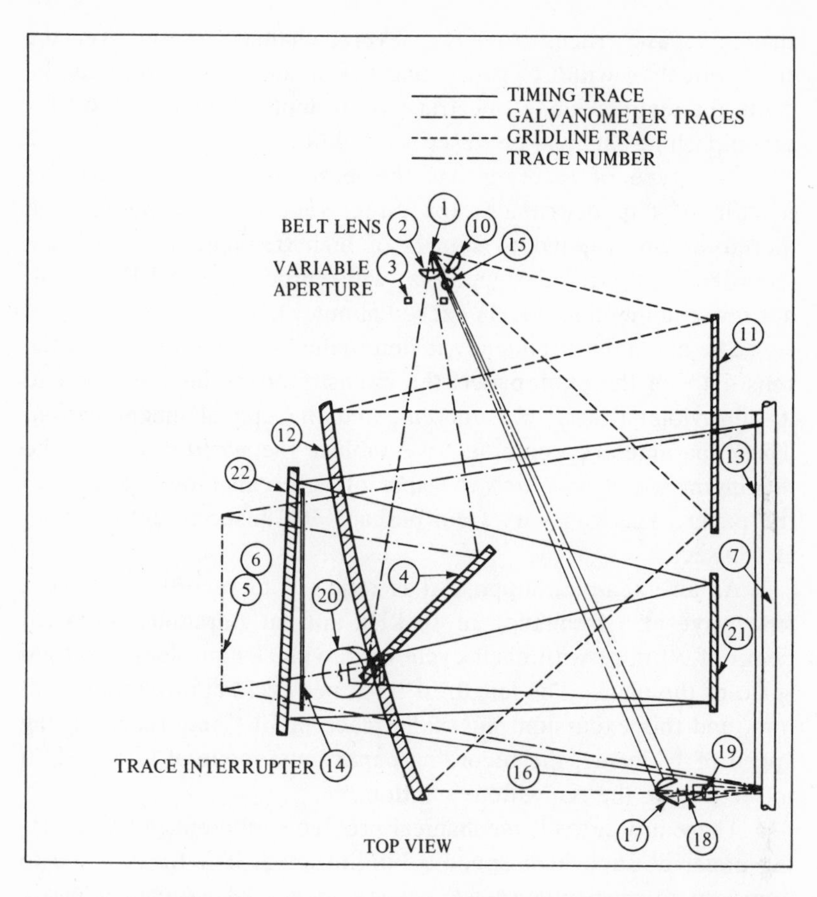
TIMING TRACE
GALVANOMETER TRACES
GRIDLINE TRACE
TRACE NUMBER
BELT LENS
VARIABLE APERTURE
TRACE INTERRUPTER
TOP VIEW

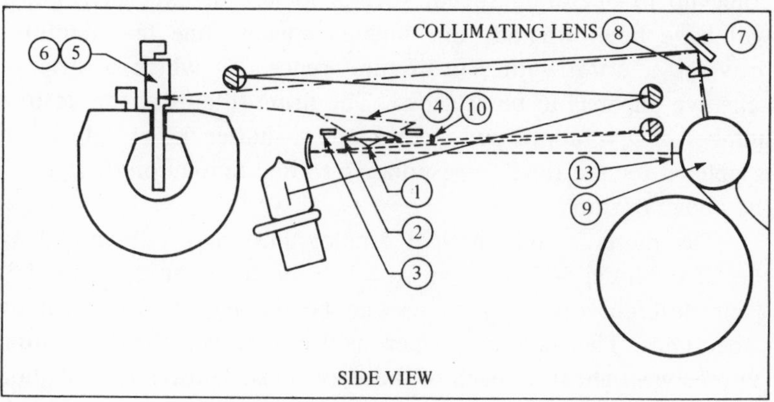
COLLIMATING LENS
SIDE VIEW

(b)

structure. The traces from the several channels each cover the entire effective width of paper, and hence some provision must be made for distinguishing one trace from another. One method for accomplishing this will be described below.

This type of recorder has the least mechanical inertia and friction of any instrument for direct recording on paper, and therefore can respond to signals of high frequency. Light-beam recorders have been designed that will reproduce a 15-kHz signal, but few commercial models exceed about 5 kHz.

The maximum frequency is determined by such factors as the sensitivity of the photopaper, the intensity of the light source and its ultraviolet ("actinic") content, and the optical magnification. These parameters combine to establish the *writing speed*, the maximum speed at which the spot of light can move relative to the paper. This can vary from perhaps 100 m/sec to greater than 10 km/sec.

As an example, suppose it is required to record faithfully a sine wave at a frequency of 10 kHz and an amplitude (peak-to-peak) of 50 mm, with each cycle occupying 1 mm along the time scale of the chart. The length of trace per cycle approximates 100 mm, and this excursion must take place in 10^{-4} sec, for a writing speed of 1 km/sec; the recording paper must move at a rate of 10 m/sec to give the requisite resolution.

There are difficult mechanical problems inherent in transporting paper at such high speeds. For instance, if it takes $\frac{1}{2}$ sec to come up to operating speed, several meters of expensive paper will have been wasted! For highest speeds, one manufacturer provides a drum with 1-m circumference, on which a strip of sensitive paper is to be attached. The drum rotates at the desired number of revolutions per second, and a shutter in the light beam is opened for the time corresponding to one revolution (0.1 sec in the above example).

The more familiar optical oscillographs may be typified by the CEC Model 5-124, shown with its optical layout in Figure 5-3. Four different optical systems can be distinguished, shown by coded lines. The recording paper, as it passes over the drive drum (9) receives light from each of these systems. Ultraviolet radiation from a quartz mercury arc lamp located at the point marked (1) is taken through lens (2) via mirror (4) to illuminate the bank of galvanometer mirrors placed along the line (5, 6). Each galvanometer reflects a portion toward the paper, onto which it is focused by a long cylindrical lens (8) located beneath the strip mirror (7). Another beam of radiation from the mercury lamp passes by way of lens (10) and mirrors (11) and (12) to the back portion of the paper roller, where it imprints on the photopaper an array of parallel grid lines corresponding to the transparent scale marks on mask (13); this scale is to assist in quantitative measurement of galvanometer deflections. To ensure correct identification of traces, an interrupter (14) is installed. This is a narrow, opaque rod moved in front of the array of galvanometers in such a way as to intercept the light momentarily from each in succession; the signal traces on the chart then show corresponding brief gaps. To further identify the traces, a serial number is imprinted along the margin. This is accomplished by the third light beam from the mercury lamp, which passes through orifice (15) and projects onto the paper an image of a numeral from a strip of microfilm (18). The microfilm is moved in synchronism with the interrupter rod so that each printed numeral appears directly opposite a gap in the trace from the correspondingly numbered galvanometer. The fourth optical system shown in Figure 5-3 takes light from a separate source, a xenon flash lamp (20), and by means of mirrors (21), (22), and (7), and lens (8), imprints a series of timing lines across the chart, one line for each flash of the lamp.

Figure 5-4 shows a record made with this oscillograph to illustrate its various features. For many applications, some of these features are not needed. Figure 5-5 shows a portion of a mass spectrogram recorded on the same type of oscillograph. The several galvanometers are connected to the same signal through amplifiers with graded gains. The zero points are offset slightly for convenience. Some maxima are easily measured on the trace of greatest sensitivity, while others that send this trace off scale can be read on less sensitive traces; all are intercomparable if the corresponding gains are taken into account. In this application, the trace identification and timing lines are not needed.

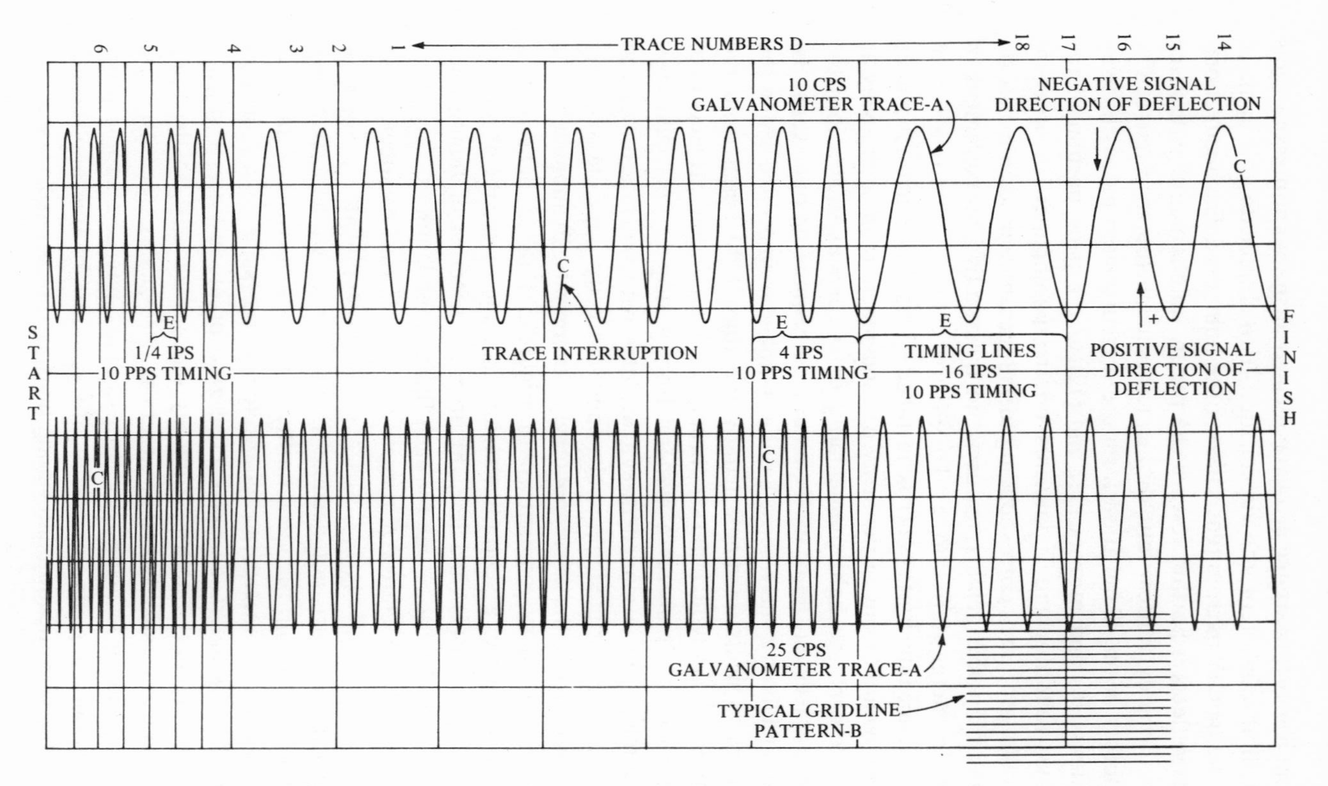

Figure 5-4. A recording from the oscillograph shown in Figure 5-3. Note the trace interruption marked C.

The recording paper most often used in this and many other oscillographs is of the type called *direct recording* or *printout paper* (POP). The intense ultraviolet light produces many latent image centers in the silver halide grains of the paper, and these are caused to grow into larger silver particles by exposure to ordinary room light. If greater speed is required, a low-intensity ultraviolet illuminator, called a *latensifier*, can be installed so as to provide the needed post-exposure to the chart as it passes out of the recorder. This "optical development" produces a high-contrast image quickly and without the use of chemical processing.

Traces on POP paper tend to fade into the background upon prolonged exposure to room light (and much shorter exposure to direct sunlight), an effect that can be prevented by optional chemical development and fixation. Chemical development has the added advantage of increasing markedly the sensitivity of the paper, so that the highest writing speeds can only be attained in this way. Manufacturers generally provide optional automatic equipment for chemical development; in some cases, this connects directly to the recorder so that the paper need not be handled or exposed to room light before development. Tray development is convenient if it is only to be used for an occasional short segment of a recording, as for publication purposes, where permanence is important.

X-Y OSCILLOGRAPHS

Several companies make X–Y recorders using a doubly deflected light beam. Figure 5-6 shows the optical diagram for one of these. The beam of radiation is first deflected in the X direction by a galvanometer responding to one variable, then by a second galvanometer operating in the Y direction in response to a second variable. Photosensitive paper in separate sheets is held against a glass screen, then developed with or without latensification in the usual way.

One manufacturer (Siemens) makes use of a 1-mW helium–

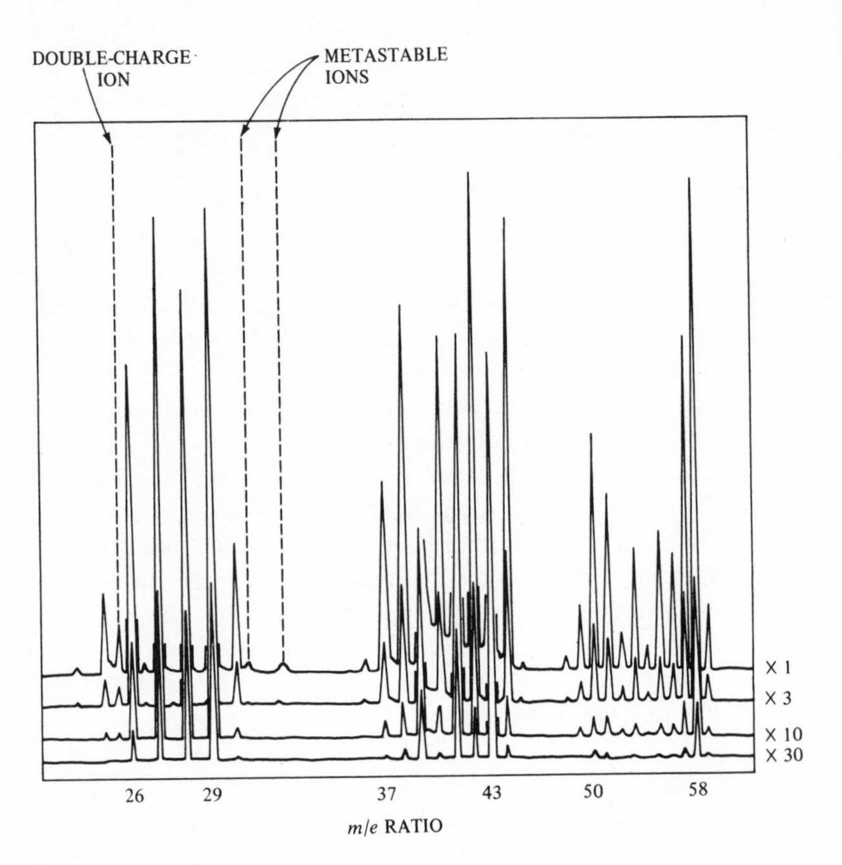

Figure 5-5. A mass spectrogram produced by the recorder of Figure 5-3.

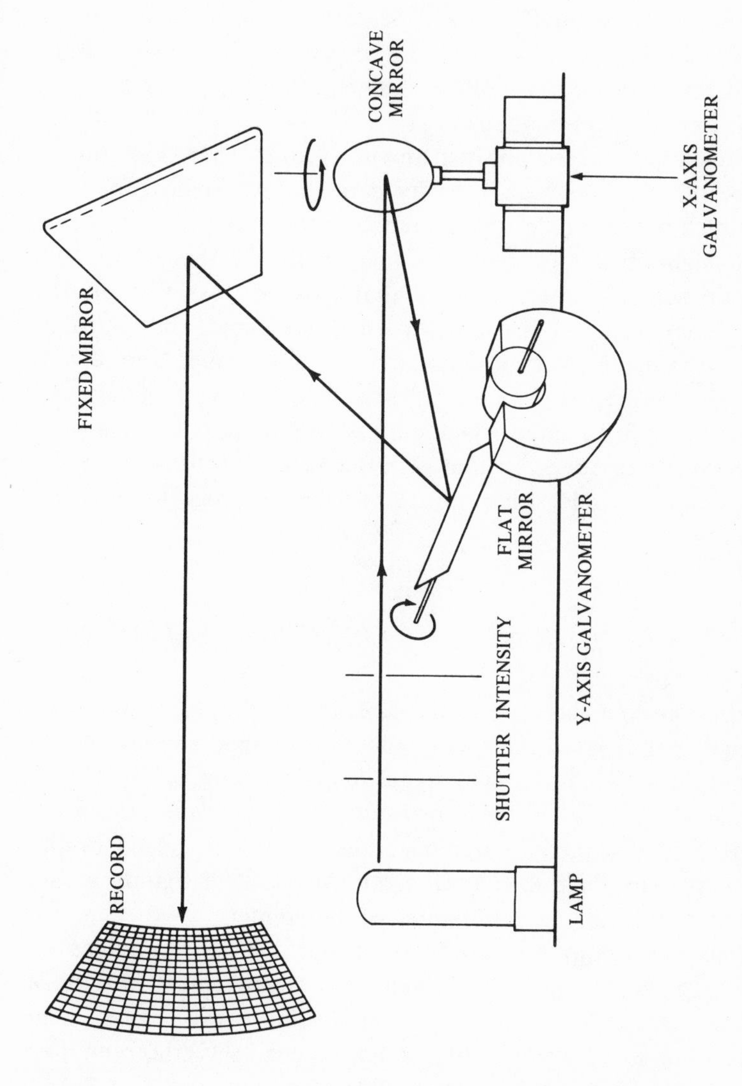

Figure 5-6. The writing mechanism for an optical *X–Y* oscillograph. The light beam is deflected by two movable mirrors (one concave, one flat) whose position corresponds to the *X* and *Y* signals.

neon laser as a light source in an X–Y oscillograph (Figure 5-7). The luminous intensity is sufficient, in spite of the low actinic quality of the red light from the laser, that a writing speed of 60 m/sec can be attained. Further versatility can be had by allowing the laser beam to emerge from the recorder and provide a trace on a distant projection screen for audience viewing. Signals up to 250 Hz can be accommodated.

Honeywell offers an instrument (Model 1806 Visicorder; Figure 5-8) that is a high-intensity cathode-ray oscilloscope with provision for convenient manipulation of recording paper over a special high-efficiency fiber-optic face plate. By this means any trace or combination of traces that can be observed on the oscilloscope can be directly recorded on paper. In addition, provision is made for continuous movement of paper from a roll, up to 250 cm/sec, to permit recording of time-variant signals. An oscilloscope can be intensity-modulated, a feature not available in conventional recorders, so that with the present instrument, X–Y–Z plots can be made, where the Z variable controls the intensity of the record.

MOVING PEN OR STYLUS OSCILLOGRAPHS

Recorders in this class are of the direct deflection type optimized for high-speed operation, with maximum paper speeds of the order of 200–500 mm/sec and signal frequencies up to 150 Hz. A mechanical deflection system to respond to such frequencies must be specially designed to minimize the mass and inertia of the moving system while keeping it rigid. Any lack of rigidity would result in distortion due to whipping of the pointer.

The maximum amplitude of deflection in this class of oscillograph is usually about 50 mm, so it is convenient to arrange multiple-channel recorders in a side-by-side array, rather than overlapping as is usual with optical types. (Overlapping pen drives would be difficult or impossible to design anyway.) A few instruments (Hewlett-Packard's Model 7402A, for example) provide an option whereby two adjacent channels can be operated

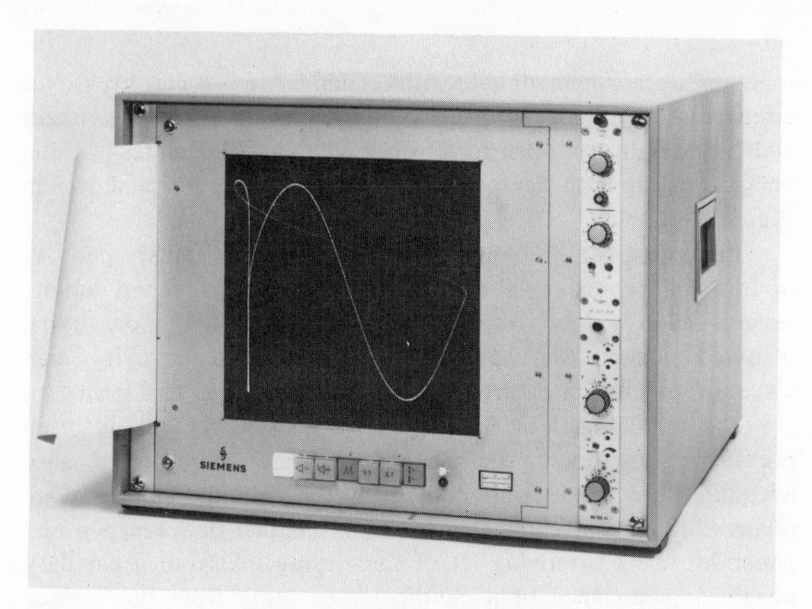

Figure 5-7. The Siemens X–Y oscillograph, which uses a laser beam as a light source.

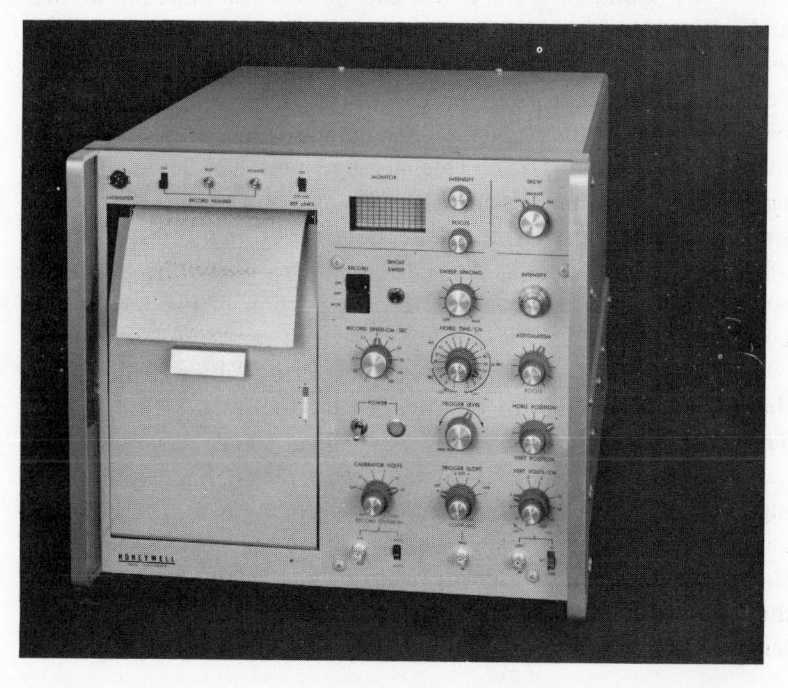

Figure 5-8. A cathode-ray oscillograph. The signal viewed on the monitor is recorded permanently on paper.

together as a single double-width channel; a special crossover circuit directs the signal to the left-hand pen for negative voltages and to the right-hand pen for positive voltages, the center position being zero for both pens (Figure 5-9). The recorder is shown in Figure 3-9(b).

Recording can be by ink supplied through a capillary pen. As in any ink recorder, a compromise must be reached among several features: low viscosity needed for easy flow, slow drying to avoid plugging the capillary during periods of inactivity, high viscosity to eliminate spreading of the trace, and fast drying to prevent smudging. For high-speed recorders, the ink is often pressurized by pneumatic means or a small pump, to ensure adequate flow at the relatively high writing speeds required. One recorder (Siemens) eliminates frictional contact between pen and paper by using a moving jet of fast-drying ink from a capillary nozzle at a distance of a centimeter or more from the surface (Figure 5-10).

Many manufacturers avoid the difficulties inherent in pen-and-ink systems at high writing speeds by substituting thermal writing. In the usual form, the deflecting pointer carries a linear heating element that forms an extension of the pointer itself. The paper is made to pass over a sharp angle where it is touched by the heater (Figure 5-11). This arrangement has several advantages. It ensures that the heater and paper make contact at only a single point; it linearizes the deflection, according to the geometry of Figure 2-3(a); and it tends to minimize a cooling-off effect, since when the signal changes, a new portion of the heater wire becomes effective.

BLH, Inc., in their oscillographs, use Peltier heating in a thermocouple junction to produce a very localized heating effect. The trace must be linearized by mechanical means.

Heat-sensitive paper is coated with a plastic material containing myriad microscopic bubbles over a black base. The unheated paper appears white, but where it contacts the heater, the plastic is melted momentarily, which destroys the bubbles, and renders the coating transparent so that the trace appears black. The temperature of the stylus is generally controllable by the operator,

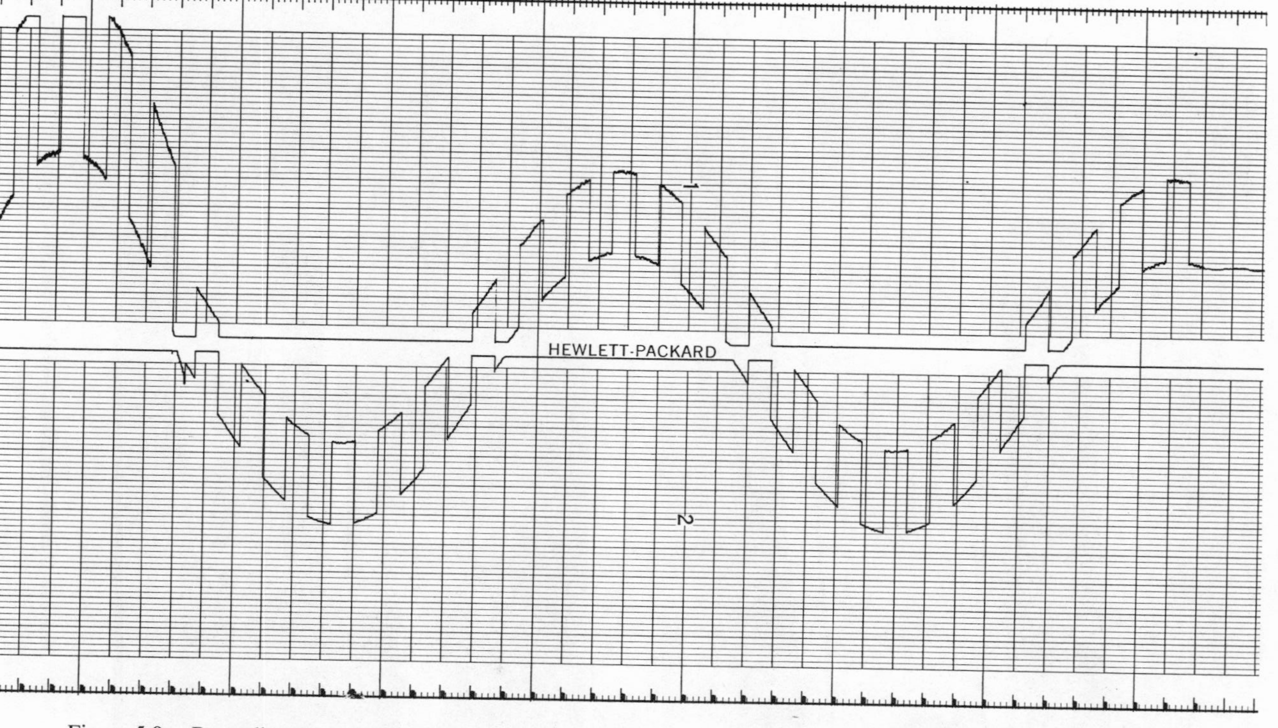

Figure 5-9. Recording from the Hewlett-Packard Model 7402A. The two channels are operated as a single channel.

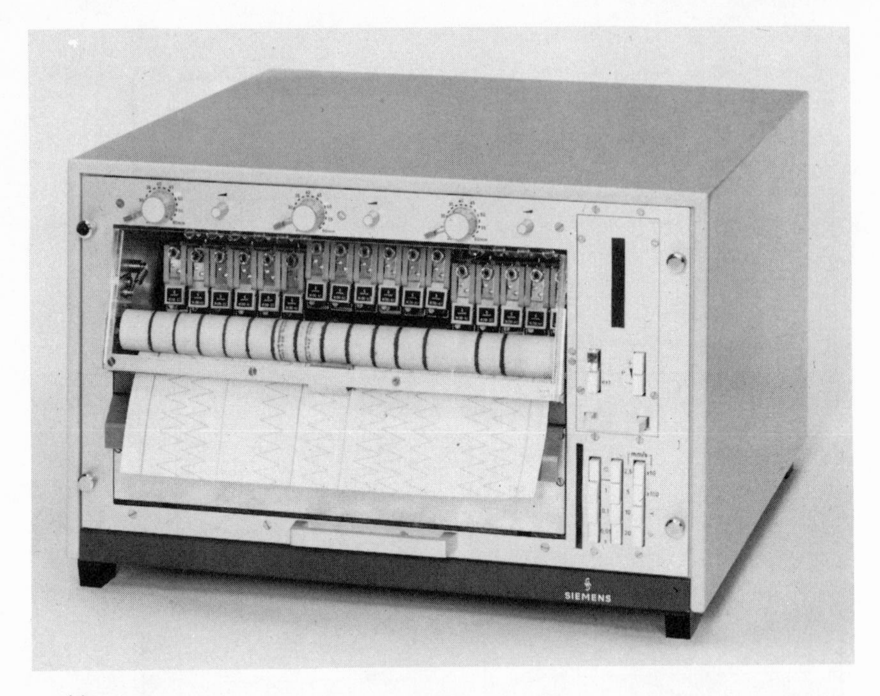

(a)

(b)

Figure 5-10. (a) The Siemens liquid-jet oscillograph. (b) Closeup of the writing mechanism.

since it must be higher for rapidly changing signals than for those that vary only slowly. It must also be raised if the paper speed is increased, and an automatic adjustment is often built in if the recorder has multiple chart speeds.

Also applicable to oscillograph recording is paper sensitive to the passage of electricity. The earliest system of this kind, no longer used, required a high voltage (~300 V) to puncture the paper with a series of sparks, burning off the surface material. This was hazardous and noisy, as well as producing objectionable fumes. A modern version (as used by Hewlett-Packard) requires a paper impregnated with zinc oxide; passage of a direct current at less than 10 V reduces the oxide to free zinc, forming a black trace. Another type (Alden) removes by electrolysis traces of the metal of which one of the electrode contacts is made (for example, iron from a steel contact bar). The metal ions then react with a constituent of the paper to form an intensely colored trace. Either of these methods will produce a sharp line of high contrast.

Electrical recording paper is more expensive than thermal, and both are more costly than paper suitable for ink writing. The electrical method is chiefly used for unattended installations, since its long-term reliability is highest.

FIXED-STYLUS RECORDING

Varian manufactures a series of oscillographic recorders under the trade name *Statos*, which achieve high writing speed by eliminating entirely all moving parts other than the paper transport mechanism (Figure 5-12). The recording head contains 100 fixed styli spread across the 100-mm effective paper width. An electrical signal applied to any stylus causes an electrostatic charge to be deposited on the coated paper, which then passes through a toner. The black particles of the toner are picked up by the charged areas to produce a clear and permanent record. (This is the well-known printing method of *xerography*.)

Figure 5-13 shows schematically how the Statos recorder works. The instrument prints its own coordinate grid, just as do

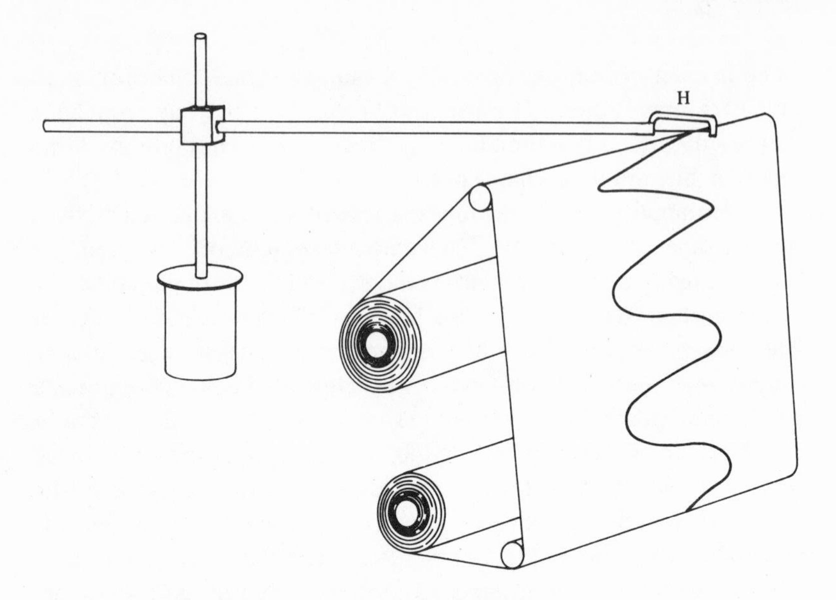

Figure 5-11. Thermal writing mechanism. The heater *H* contacts the paper at a single point determined by the sharp angle in the paper.

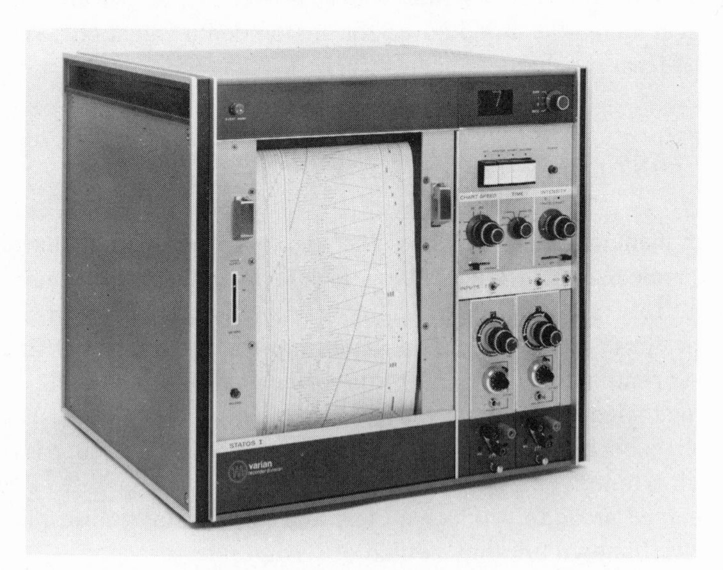

Figure 5-12. Statos (Varian) recorder. Writing is accomplished with fixed stylus and xerography.

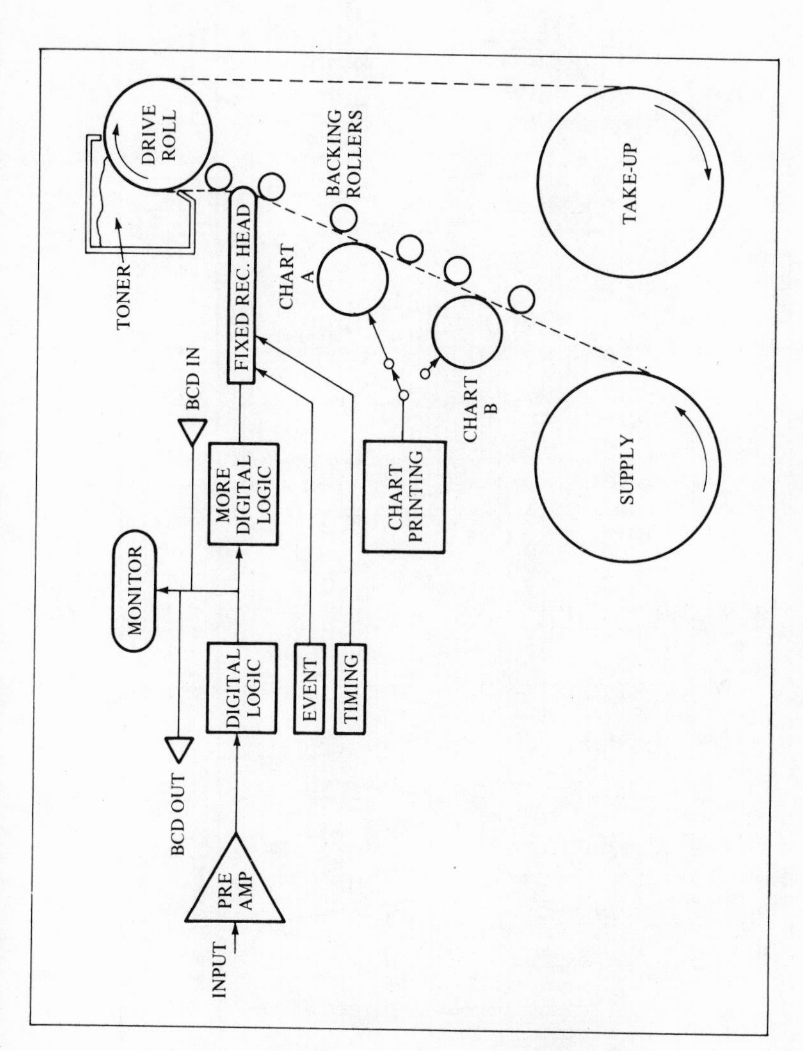

Figure 5-13. Block diagram of the Statos recorder.

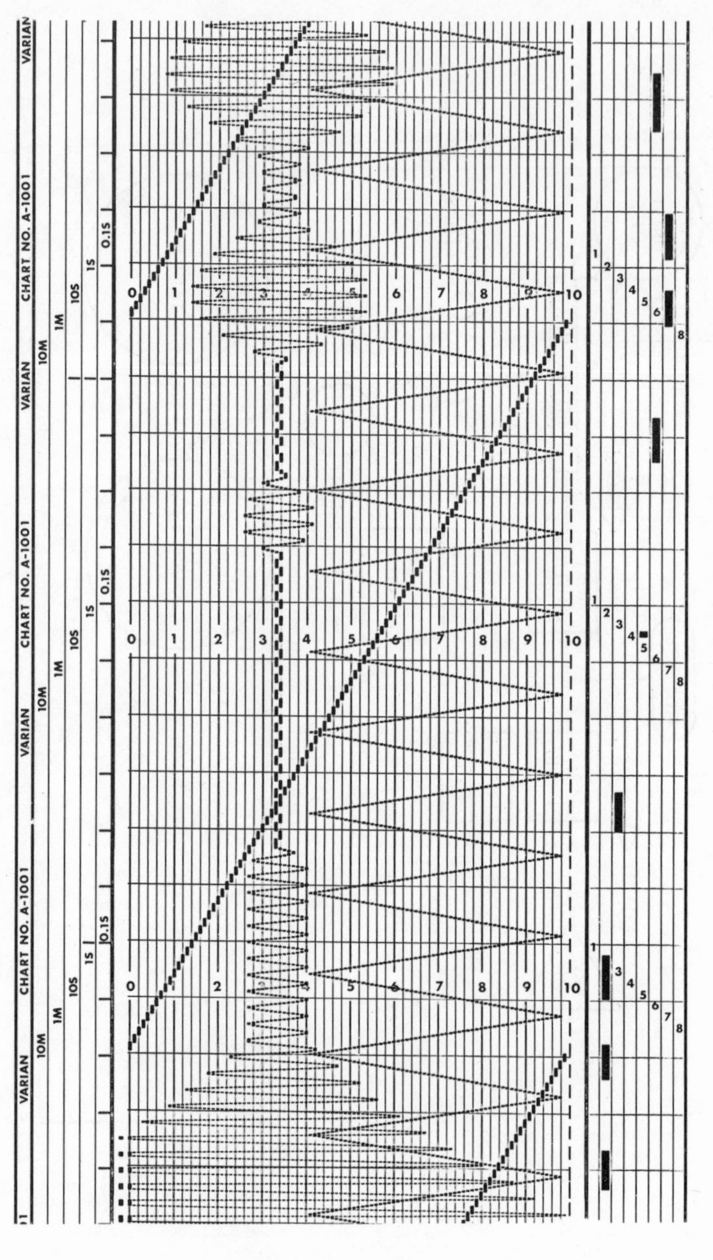

Figure 5-14. Sample recording produced by the *Statos* recorder.

many photographic oscillographs. Either of two rollers (designated *A* or *B*), when energized, imprints electrostatic charges appropriate to a selected grid pattern. Timing lines are imprinted through the same recording head that carries the signals, ensuring no offset between the two.

The diagram shows two input channels, one for the usual analog signals (via a preamplifier), the other to accept a signal in digital form (marked "BCD in"). Other models have various combinations of analog and digital channels. The analog signal must be converted to its digital equivalent in order to give ON or OFF commands to the individual styli. Several channels for event timing records are provided along one edge of the paper. A sample recording is shown in Figure 5-14. It is evident that all traces consist of a series of dots. This instrument, because of its digital logic, is particularly versatile with respect to interfacing with a computer.

A similar multistylus recorder, using electrical rather than xerographic writing, was formerly manufactured by General Radio (Model 1520). It was designed especially for recording low-frequency transients.

6

Paper Feed and Writing Mechanisms

PAPER FORMS

The majority of analog recorders utilize either strip charts or circular charts. The strip charts, supplied in rolls, have the great advantage that they can operate continuously for long periods without need of replacement. Circular charts, on the other hand, must be replaced periodically, and hence are used mostly for recording slowly changing variables, a single chart to last, perhaps, one day or one week (Figure 6-1).

Roll charts are inconvenient from the point of view of storage and retrieval of information. If only short segments are of permanent interest, as is often the case in connection with laboratory experiments, the strip must be cut up, and the useful portions filed for reference. The paper tends to retain its curl, which adds to its inconvenience. To overcome this failing, fan-fold (or ''Z-fold'') paper has been introduced. In this form, the supply of paper is stacked flat, is pulled through the feed mechanism, then stacks itself flat again (Figure 6-2). The fan-fold

Figure 6-1. A variety of recording papers and ink (Graphic Controls Corporation).

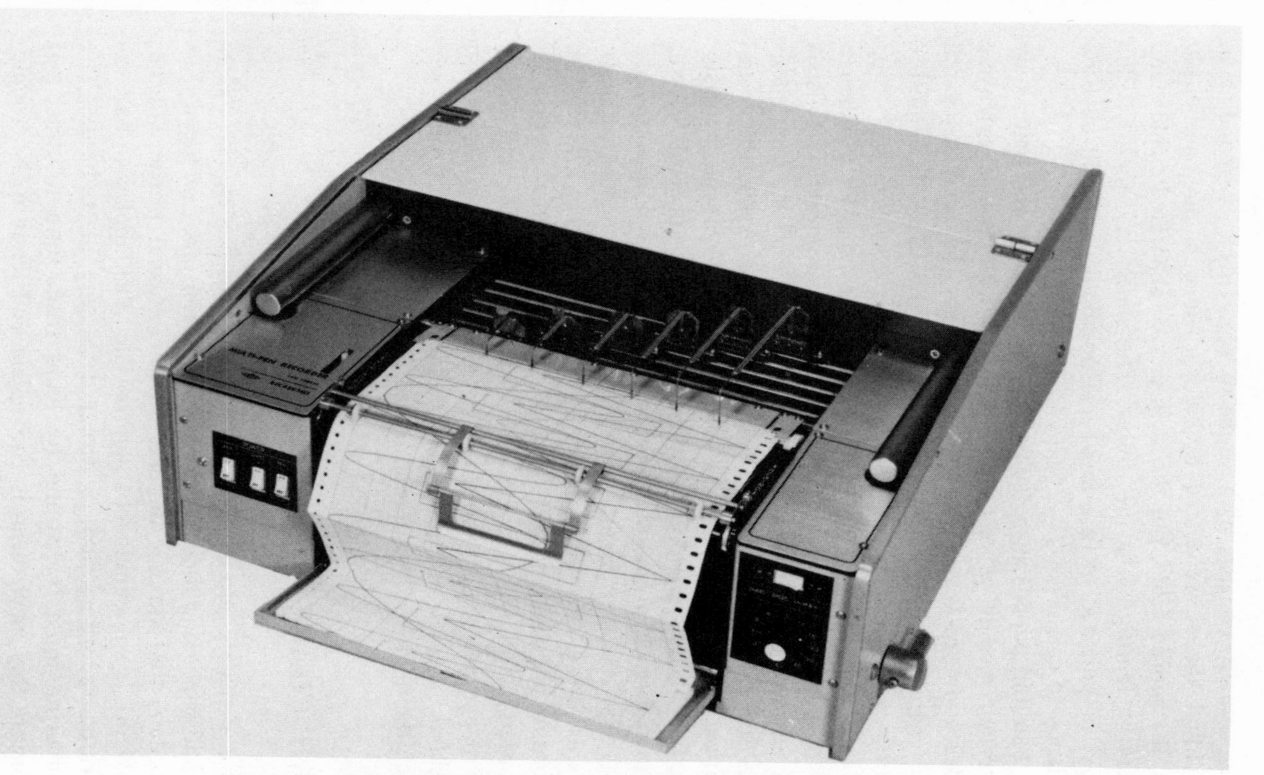

Figure 6-2. Z-fold paper used in an overlapping multipen recorder (Soltec Co.).

paper actually consists of a series of segments corresponding to individual sheets, separated by folds. The pen can pass easily over the scored folds, but the strip can be separated readily into sheets by tearing along the scores. The sheets can then be filed easily, or punched to fit a loose-leaf notebook, and will have no tendency to curl.

Fan-fold paper can be substituted directly for roll paper only in certain models of recorders; others would require extensive modification. In general, it can be said that recorders using roll paper can be more compact than those designed for fan-fold paper.

An alternative particularly adapted to a recorder built in as part of a larger instrument for special-purpose operation consists of a series of printed sheets connected end to end in a continuous roll. Some means must be provided for adjusting the initial position of the paper so that the start of a recording will register exactly with the zero point of the printed form. Successive forms are then torn off at the conclusion of their respective experiments.

To go one step further is to supply paper in separate sheets. This is done in many laboratory instruments with built-in recording mechanisms. It also is standard practice with X–Y recorders. A new sheet must be inserted for each experiment. The paper can be conveniently preprinted and prepunched for notebook binders. The only disadvantage is the limited time over which the instrument can operate unattended.

STRIP-CHART DRIVE MECHANISMS

The essential requirement of a mechanism for feeding the roll or fan-fold paper of a time-based recorder is constancy of speed. For recorders that do not depend on an external power source, clockwork is sometimes most appropriate. This applies to battery-operated recorders and to direct deflection instruments. The mainspring must be wound by hand from time to time, but otherwise this mechanism is convenient and likely to be trouble-free.

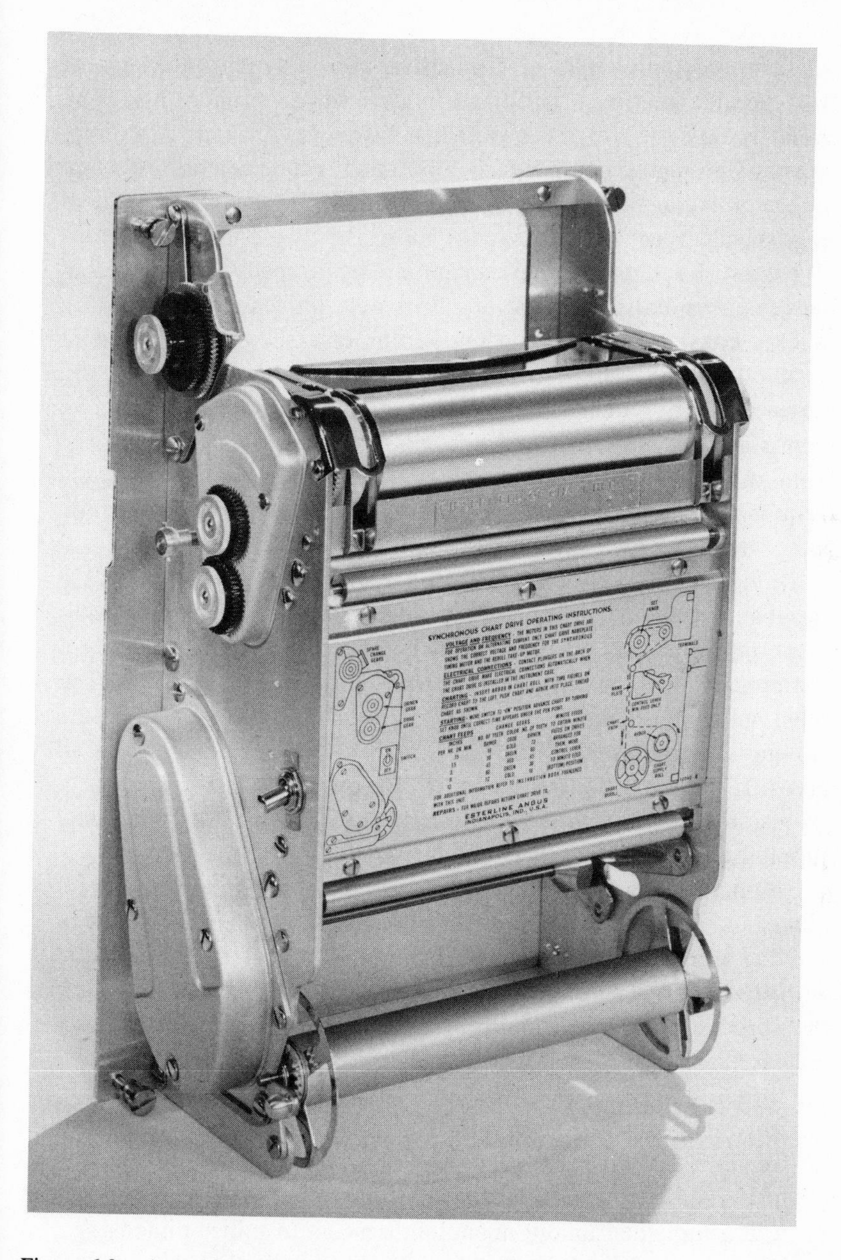

Figure 6-3. A synchronous motor chart drive. Note the additional set of gears used for changing chart speed (Esterline-Angus).

The most prevalent electric drive uses a synchronous motor with suitable gearing to produce the desired rate (Figure 6-3). The speed is thus tied to the power-line frequency, which is entirely constant enough for nearly all laboratory requirements. A stepping motor can be used in this application. It may be thought of as a special type of synchronous motor that responds to a series of pulses. Each pulse causes rotation by a specified number of degrees, typically 1.8° or 5°. Between pulses, these motors quickly coast to a stop (in a few tenths of a second). If powered with 60-Hz ac, the motor will turn continuously with the corresponding synchronous speed, 18 or 50 rpm for the above examples. The advantage of this motor is that it can be powered from a solid-state pulse generator (multivibrator) of variable frequency; changing speeds becomes a matter of turning a dial, rather than shifting gears.

For some *X–Y* recorders, the paper must be driven at a variable speed and even be reversible in direction. The mechanism is then driven by a servo system, often identical in components with the pen-drive servo of the same recorder. The drive must have power enough to overcome the inertia of the moving system, forcing it to conform with the input signals at all times. If this recorder is to be used optionally against a time base, provision must be made for bypassing the control amplifier and supplying a constant voltage to the meter.

For any type of drive to a strip chart, the final transfer of power is from a roller to the paper. Except for high-speed oscillographs, this step always makes use of a sprocket integral with the roller. The paper is perforated along the edges to engage the sprocket teeth. The perforations along one side of the paper are circular, fitting snugly over the hemispherical teeth, while at the other margin, the holes are usually elongated, to ensure correct register at the first side, but provide an additional degree of freedom laterally to allow for possible dimensional change (Figure 6-4). It is most unfortunate that little standardization has been accomplished among manufacturers as to paper dimensions.

The paper speed in oscillographs is generally too great to permit use of the sprocket and hole device without tearing the

Figure 6-4. Strip-chart paper being loaded. Note the circular holes on one side, elongated holes on the other (Leeds and Northrup Co.).

paper. Friction between driven rollers and paper is substituted. In those oscillographs that print their own coordinate lines, precise control of paper speed is not as urgent as with those using preprinted paper.

Many roll chart recorders are provided with takeup spools. These must be permitted to adjust their speed to that of the paper, and hence must be fitted with a slip clutch. For laboratory purposes, the takeup reel is often bypassed, the paper being allowed to flow out from the recorder, to be cut off as needed.

In most general-purpose strip-chart recorders, provision is made for changing the paper speed to meet the requirements of various applications. The most common speed changer operates through variable gear ratios (Figure 6-3). In some, gears must be interchanged manually, a process that will take some minutes, and hence cannot well be done while a recording is in progress. Others incorporate a convenient gear shift operated by a lever (Figure 6-5). Some designers choose to install two synchronous motors linked through a complex differential gear mechanism to give several speeds according to which motor (or both) is energized. Still others use small clock-type motors with integral gearing, which can easily be replaced as a unit. The use of a stepping motor for variable-speed operation was mentioned above.

SEPARATE SHEET RECORDING PAPER

In X–Y and other recorders that use individual sheets, several mounting arrangements have been devised. In some, the paper is placed on the convex surface of a drum and held in place by clips (Figure 6-6). The drum then rotates at constant or controlled speed, the analog of the independent variable, while the pen moves along an arm parallel to the axis of the cylinder. This is relatively convenient to design and to use, but only part of the record is visible to the operator.

Other models use flat bed paper mounting. Means must be provided for holding the paper firmly in contact with the

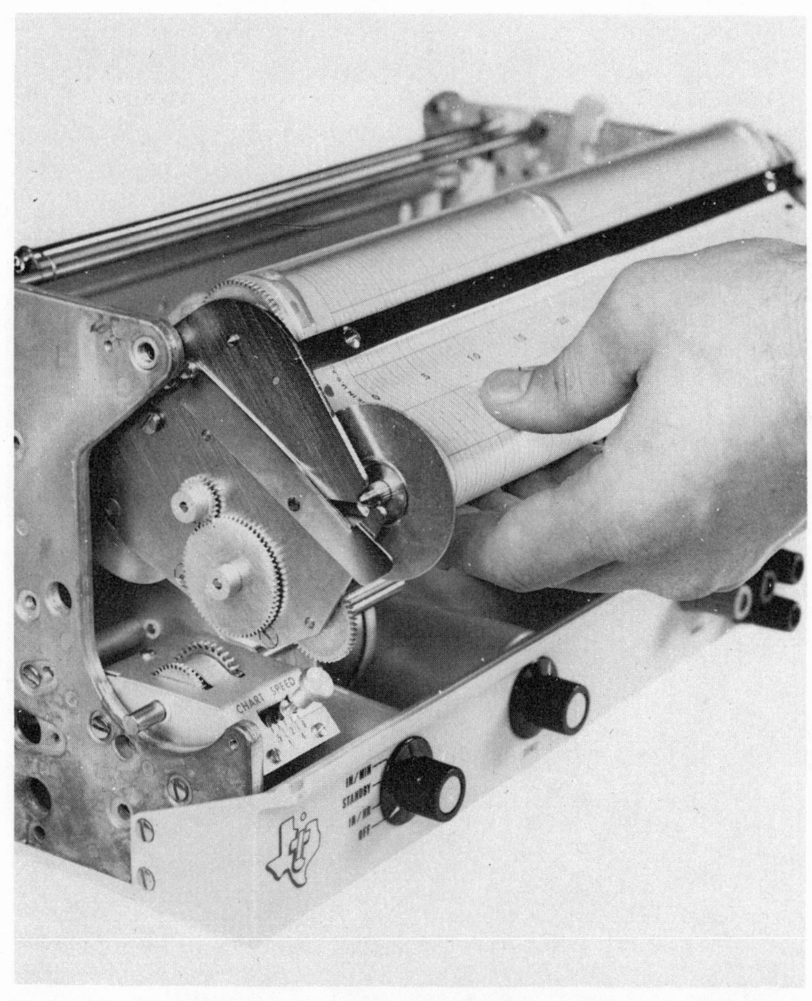

Figure 6-5. Variable paper speed is achieved with a simple gear shift lever (Texas Instruments, Inc.).

Figure 6-6. *X–Y* recorder of the drum type in a laboratory instrument (Perkin-Elmer).

supporting platen, a need that does not arise in strip-chart recorders, where the paper is under tension. The paper is sometimes held down by mechanical or magnetic clips. Another widely used method is by air pressure; the platen is pierced by many small holes, and a fan is located behind it to produce suction, so that atmospheric pressure holds the paper in place. At least one major manufacturer uses an electrical hold-down; high-voltage dc is applied to the insulated metal platen, and electrostatic attraction holds the paper in place. One manufacturer has tried a magnetic platen; the paper is printed with magnetic ink, and the resulting forces are adequate to immobilize the paper.

Mention should be made in this context of the special-purpose recorder by Leeds and Northrup designed to draw a graph on a 5 × 7 in. Keysort card (Figure 6-7). A blank card is fed in manually until it is caught up by moving rollers, then the recording experiment is started, at the close of which the card with its record emerges.

PRINTING MECHANISMS

Capillary Pens

By far the majority of laboratory recorders use ink fed through a capillary pen. This has an inherent defect in that the fast-drying ink necessary to prevent smudging of the record tends to dry out in the pen, clogging it. Many attempts have been made to develop an inking system without this difficulty, and many have met with at least partial success. One possibility is to use paper with some degree of absorbance, so that the ink will penetrate immediately, and not require the quick-dry feature. This, however, tends to make the paper sensitive to changes of dimensions with ambient humidity.

Another possibility is to pump the ink under pressure, which is an effective way to keep the capillary from clogging. This is likely to eject too much ink, particularly at slower writing speeds. It has marked advantages at high speeds, however, in preventing skipped areas, and so is widely used in oscillographs. The pump

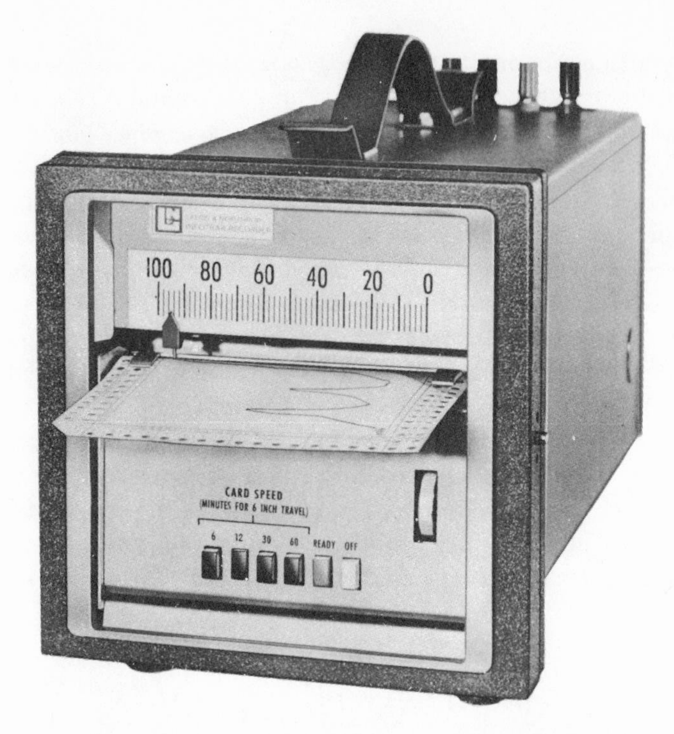

Figure 6-7. Leeds and Northrup instrument which records on a 5 × 7 in. card.

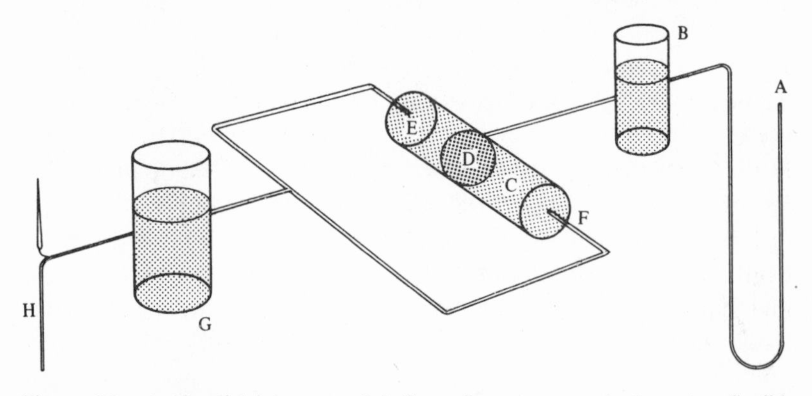

Figure 6-8. An inertial ink pump. Ink flows from a reservoir through a flexible cable (A) to a small reservoir (B). There is a steel ball (D) inside the cylinder (C), which has a valve on either end (E and F). Ink is forced through one valve or the other when the pen carriage on which B and C are located moves in response to a signal. Consequently the pen reservoir (G) and pen (H) receive ink at a greater rate when the pen moves (accelerates). (Esterline-Angus.)

may take a form that delivers ink in accordance with the speed at which the pen motor is responding, so that ink will be made available at a greater rate when the pen is moving faster (Figure 6-8). In other recorders, fast-drying ink is pumped continuously, forming a jet from capillary tip to paper, a distance of perhaps 1 cm, so that friction between pen and paper is eliminated.

The ink supply for a nonpumping capillary pen is sometimes held in a refillable reservoir, which may move integrally with the pen (Figure 6-9), or may be stationary, feeding ink through a flexible tube (Figure 6-10). Alternatively, the ink may be supplied in disposable cartridges that fit directly onto the pen carriage (Figure 6-11). In any of these possibilities, the ink is transported to the capillary tip by a combination of gravity and surface tension (capillarity) forces.

Other Mechanisms

A number of recorders have been designed to use commercial fountain or ball-point pens, but these have not enjoyed wide popularity. Drafting pens ("LeRoi," "Rapidograph," etc.) have been used occasionally, with fair success. Felt-tip pens are perhaps as reliable as any (Figure 6-12).

Many designers have abandoned ink altogether, choosing instead to use paper sensitive to pressure, heat, electricity, or light. All of these except the pressure-sensitive type are used most extensively in oscillographs, and have been described in Chapter 5; hence they will not be discussed here in detail.

Pressure-sensitive paper is used primarily in low-speed miniature recorders. It consists of black paper coated with a wax or plastic material, similar to heat-sensitive paper in containing microscopic bubbles of air that render it opaque. Pressure destroys the bubbles so that the black backing shows through. This paper is used with a deflection instrument equipped with an impact bar. At regular intervals, such as twice a second, the bar presses momentarily against the meter pointer, causing it to make a dot on the chart. The analog recording is thus a series of fine dots, close enough usually to give the appearance of a continuous

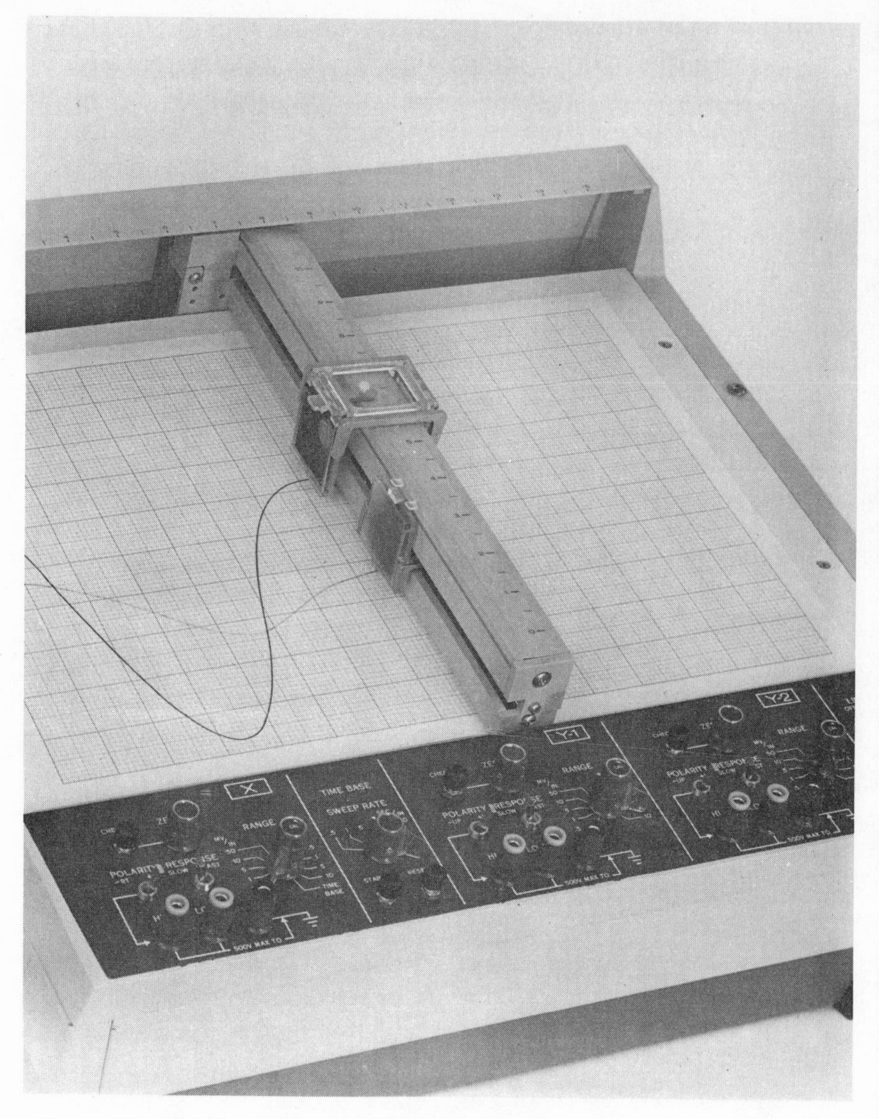

Figure 6-9. Capillary pen with ink source that moves with pen (Hewlett-Packard).

Figure 6-10. The pen is fed through a capillary from a remote reservoir that remains fixed.

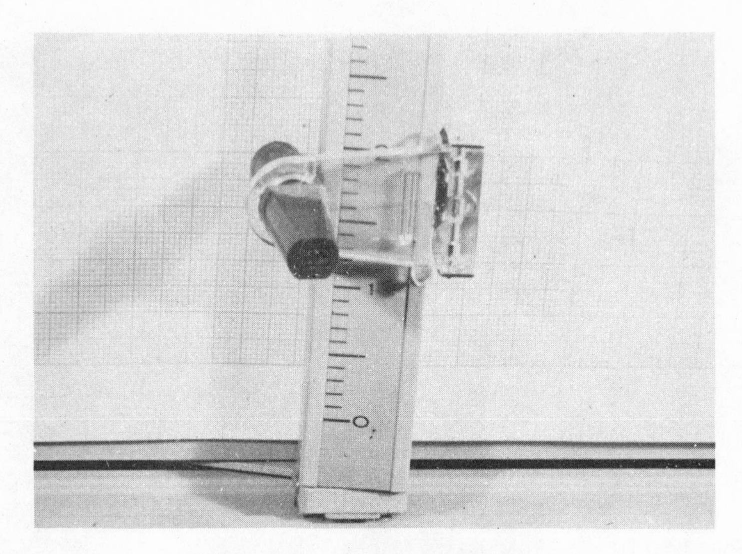

Figure 6-11. Disposable cartridge pen.

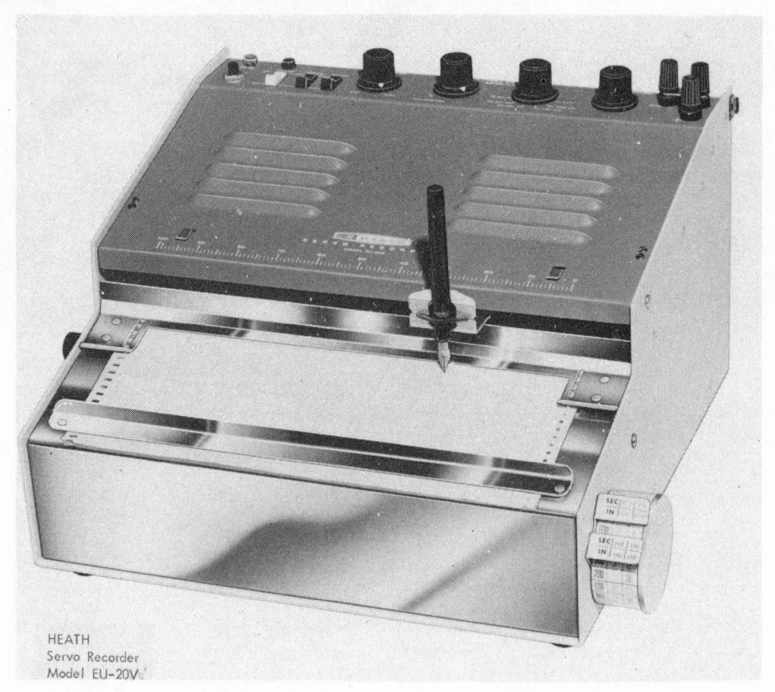

Figure 6-12. Fountain pen used as the writing instrument (Heath Co.).

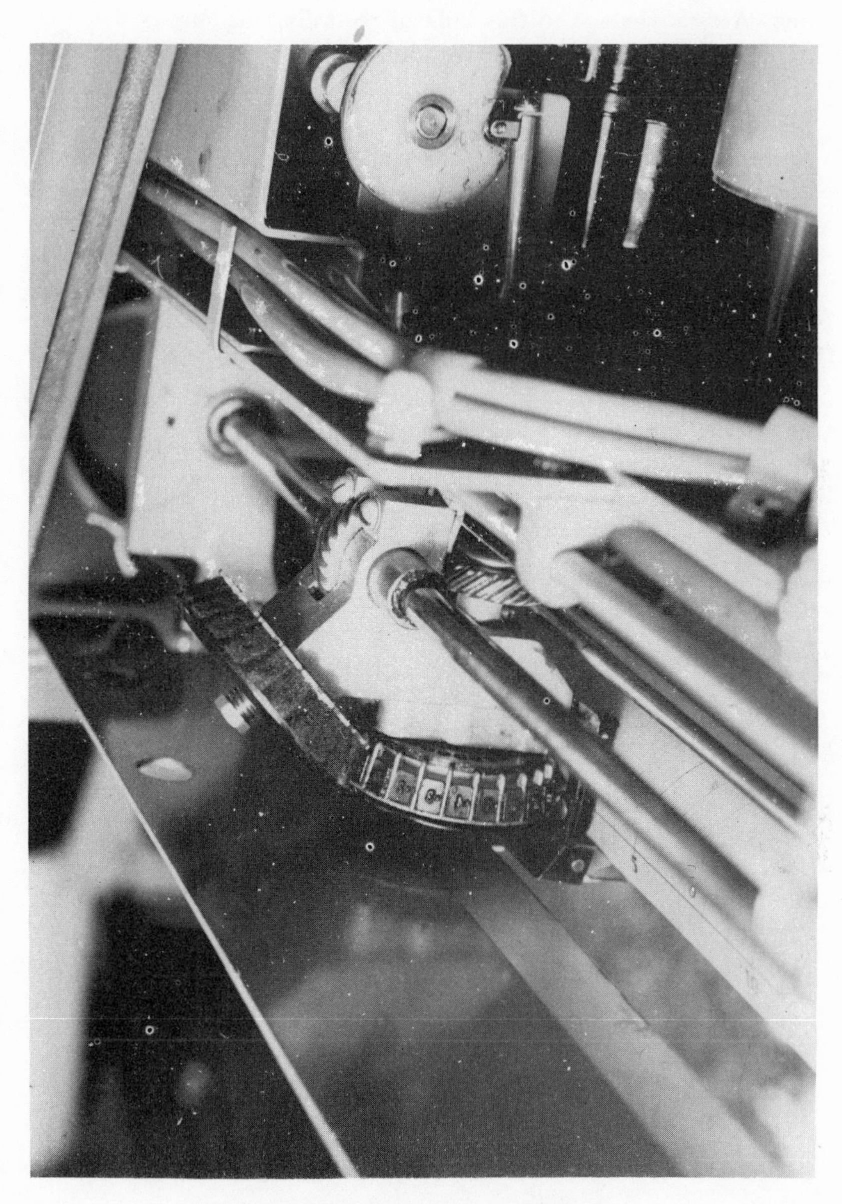

Figure 6-13. Print wheel mechanism.

line. A disadvantage to this kind of recording medium is that the paper retains its sensitivity indefinitely, hence is subject to disfigurement from the pressures of ordinary handling.

Some recorders, usually comparatively slow potentiometric types, are provided with a print wheel. This is a small, rotating device with rubber type to imprint a dot accompanied by a digit on the chart paper. It turns from one digit to the next in synchronism with a selector switch that samples successively a series of inputs. This arrangement is especially appropriate for recording on one chart the potentials of a family of thermocouples associated with different parts of an apparatus. Each thermocouple is responsible for a series of dots with a corresponding printed number (Figure 6-13).

7

Shielding and Grounding

Modern recording instruments frequently have sensitive preamplifiers built into their circuitry or available as optional plug-in equipment. Sensitivities of 0.5 mV/inch are common and sensitivities as high as 1 μV/inch are available. The use of sophisticated electronic amplifiers in recorders greatly increases their capability but at the same time requires increased sophistication on the part of the user to deal with the problem of electrical noise.

INTERNAL NOISE

Noise problems can conveniently be divided into two classes of noise source—internal and external. By internal noise we mean effects such as shot, flicker, Johnson, and inherent amplifier noise. External noise encompasses effects such as ground loops, ac pickup, leakage current and thermoelectric voltages.

Shot noise (also called Schottky noise) is a characteristic of semiconductor devices which results from fluctuations in the number of carriers conducting current through the device. Flicker noise (also called $1/f$ noise) is less well understood but appears to be due to surface effects in semiconducting crystals.

Both contribute to the noise output of any solid-state amplifier, and the magnitude of the noise output depends strongly on temperature and frequency. These effects, together with the noise inherent in the amplifier design, will place a lower limit on the smallest signal that can be detected. We must assume that this noise is small compared to the signals of interest since if this is not the case, the only practical remedy is to buy a better amplifier.

Johnson noise affects the performance of extremely sensitive amplifiers, and may also be a factor in the noise level of the input signal. Johnson noise, also called thermal or resistance noise, is due to the random motion of molecules and charges whose energy of motion is directly related to the temperature of the material. The noise voltage due to this motion is given by

$$V_{\mathrm{rms}} = [cT(\Delta f)R]^{1/2}$$

where c is a constant, T is the absolute temperature, Δf is approximately the bandwidth of the measuring device, and R is the resistance of the device being measured. Some noise curves representing the minimum thermal noise to be expected are shown in Figure 7-1. Carbon or metal oxide resistors will produce noise levels higher than will wire-wound resistors. Because there are so many other, larger sources of noise, it is rare that a measurement is limited by Johnson noise in the source resistance. Nonetheless, it is good practice to keep source resistances as low as possible not only from the standpoint of improving signal-to-noise ratios, but for the purposes of impedance matching.

EXTERNAL NOISE

Thermoelectric voltage is a possible source of noise and can be considered external since it is usually possible to reduce thermal effects if they become a problem. Normally thermals give difficulty only when working with signals that are on the order of 100 μV or smaller. Thermal voltages are caused by unintentional thermocouples in the input circuit of the measuring device. When

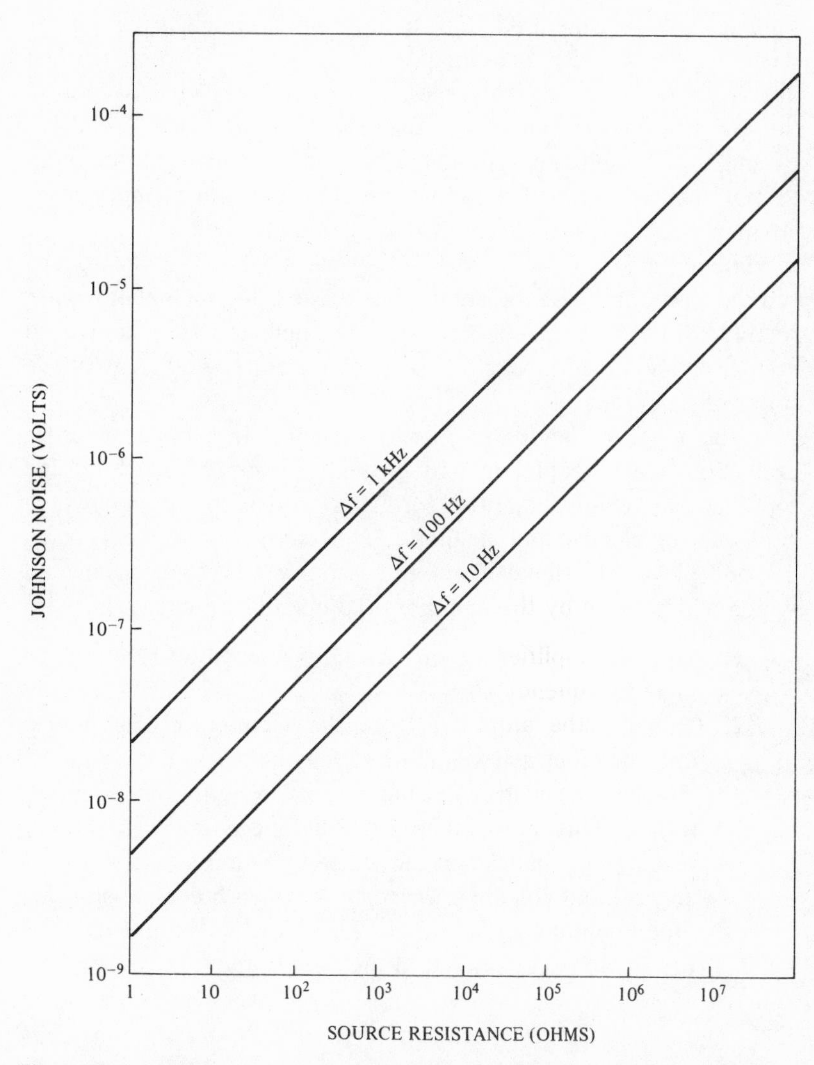

Figure 7-1. Johnson noise as a function of source resistance. Δf is the approximate bandwidth of the amplifier. (After "Electrometer Measurements," Keithley Instruments, 1972.)

two different metals are joined and when the temperature may vary from one part of the input circuit to another, thermal voltages will be generated. For example, in a circuit using copper wires soldered with ordinary lead–tin solder, a temperature difference of a degree centigrade can generate a few microvolts of thermal EMF. One can reduce thermals by using special solders, by making all leads and contacts out of copper, or by attempting to maintain all points of the input circuit at the same temperature. "Noisy" contacts which are cured by increased contact pressure or cleaning may be due in part to thermal noise. (Another way to reduce contact noise is shown in Figure 3-4.) Copper oxide readily forms on bare copper wires or contacts and a copper to copper–oxide junction can cause thermoelectric voltages on the order of 1 mV/deg.

The most bothersome sources of noise in typical recorder applications are ac pickup and ground loops. The wires of the input circuit to any amplifier form a closed loop through which time-varying electric and magnetic fields can pass. Whenever such a loop "picks up" signals from an ac source, a voltage is induced which will be seen by the amplifier. To eliminate pickup, one can:

1. Use an amplifier which strongly rejects 60-Hz and all higher frequency signals.
2. Remove the amplifier from the vicinity of any power lines or other sources of ac signals.
3. Arrange the source circuitry so as to eliminate inductive loops. This may mean twisting input leads together, placing the instrument in close proximity to the signal source, and shielding the input leads with copper braid or other conducting material. The electrical connection of the shield is considered in the next section.

GROUNDING AND GUARDING

While the practice of shielding every cable that carries a small signal or that is subjected to large interference is quite common,

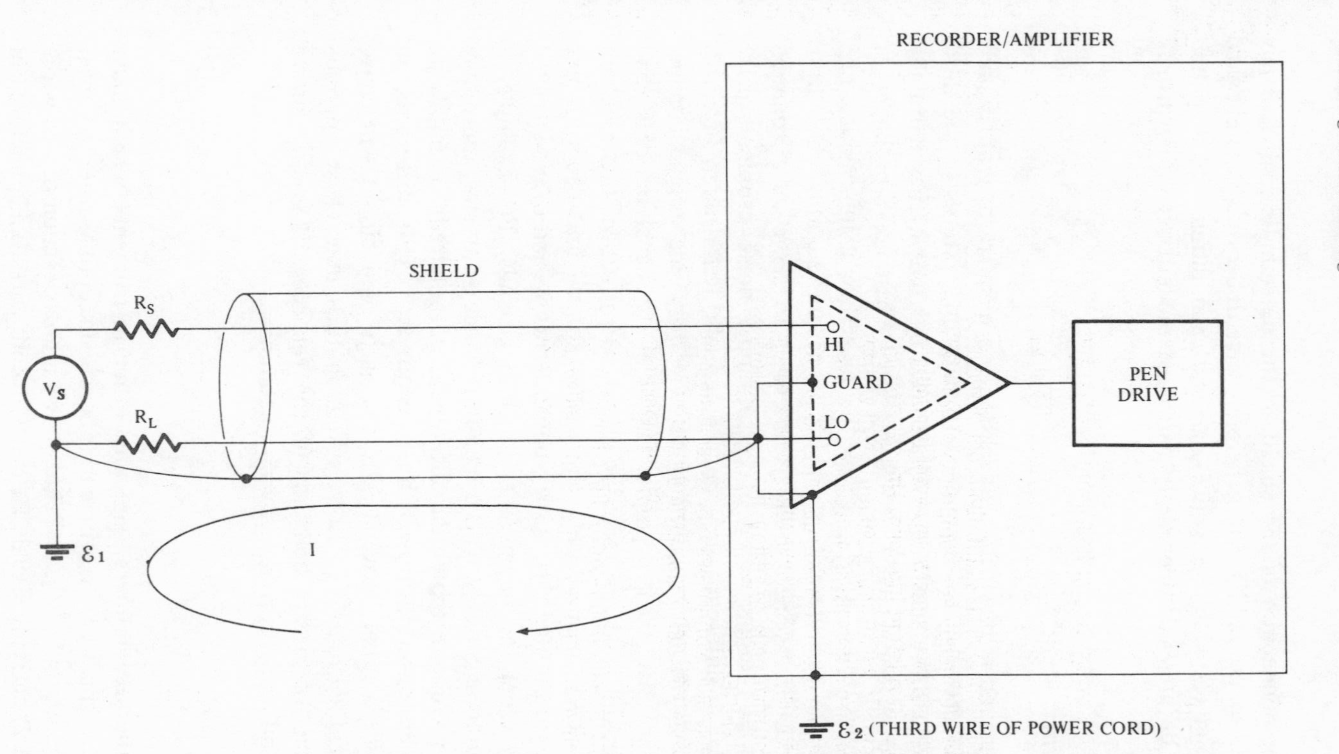

Figure 7-2. The "classic" ground loop due to improper connections. The signal originates from a well-grounded source; the amplifier is guarded and grounded using a three-prong plug. The common-mode voltage $\mathcal{E}_2 - \mathcal{E}_1$ (the difference in ground potentials) will cause ground-loop currents as shown.

proper connection of the shield to the detecting circuit and the circuit to ground is not generally well understood. The exact meaning of shield (or guard), ground, and differential amplifier must be grasped to properly use most modern recording instruments.

Ground

The ground is most often understood to mean earth ground such as that provided by connection to plumbing fixtures or the third wire of a power cord. Relay racks and instrument chassis usually should be connected to earth ground. When noise problems of unknown origin arise, perhaps the most useful information one can have is where the circuit is grounded and via which connections. While ground is intended to serve as a common potential (voltage equal to zero), many noise problems arise because of differences in ground potentials. Referring to Figure 7-2, which will be explained in detail shortly, the ground connections are made at two points. One is at the signal source V_s and the second is at the recorder chassis. In practice, this situation may easily arise if both the signal source and recorder are grounded through three-prong power cords. $\mathcal{E}_2$ and $\mathcal{E}_1$ are intended to represent the potential at the two grounds. For example, if measured accurately with a high-impedance voltmeter, there may exist a voltage (potential difference) of hundreds of millivolts between different power line "grounds." This difference in ground potential between different parts of a circuit will cause "ground currents" to flow, which in turn may cause spurious readings or erratic behavior of the recorder. It is the elusive "ground loop" current we wish to eliminate.

Guard

Recorders, amplifiers, and any other instruments which must measure small voltages or currents are susceptible to errors from leakage currents. Figure 7-3(a) shows a cross section of two wires (1 and 2) carrying signals V_1 and V_2, both wires being enclosed in a metal shield S which is grounded. The resistances R_1 and R_2

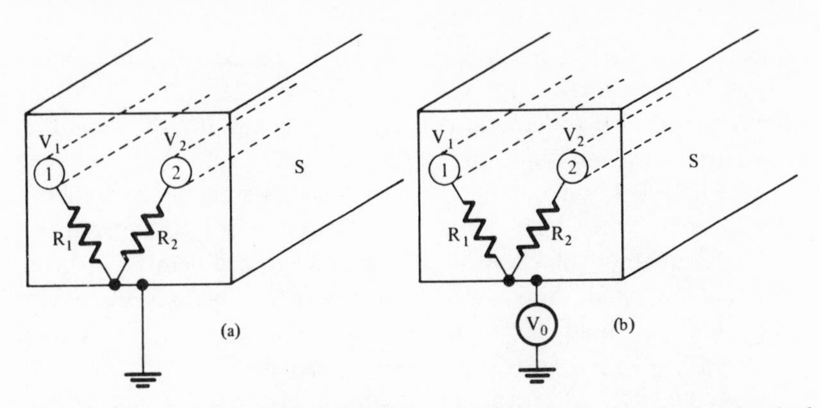

Figure 7-3. (a) A box used to shield two conductors. $V_1 - V_2$ is the signal of interest and R_1 and R_2 are the isolation resistances. (b) The difference $V_1 - V_2$ is unaffected by the potential V_0 of the shield with respect to ground. If $V_0 = V_2$, then leakage currents will be small.

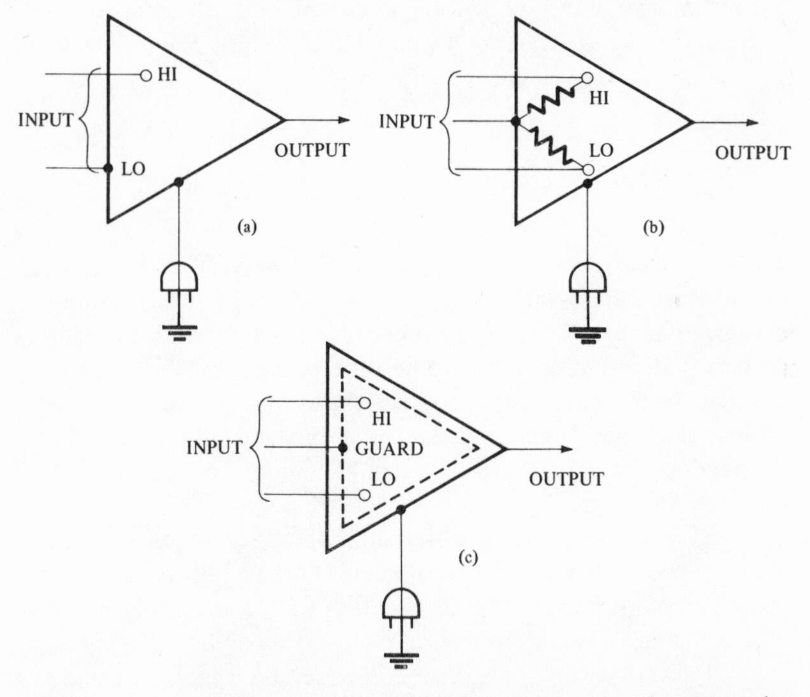

Figure 7-4. (a) Single-ended, grounded amplifier. The Lo input is connected to ground and the chassis. (b) Differential input balanced to ground. Amplifier responds to the difference between Hi and Lo. (c) Guarded differential amplifier. The dashed line represents a metal shield between the chassis (ground) and the circuitry.

may be regarded as insulation resistances and ideally are infinite. However, in practice, R_1 and R_2 are usually large but finite and consequently cause leakage currents to ground. It is obvious that once such currents flow, the observed voltage difference between wires 1 and 2 will change. Such currents can be particularly troublesome when $V_1 - V_2$ is small (1 mV or less, for example).

One method of eliminating this leakage problem is to make the metallic shield into a guard, as shown in Figure 7-3(b). Here the shield is held above ground by a potential V_0 (a shield "driven" to V_0). If V_0 is approximately equal to V_1 or V_2, then the leakage currents will be greatly reduced. The essence of guarding, then, is to surround all critical parts of the input cable and amplifier circuitry with a metal shield which is maintained at a voltage level near that of the signal being measured. In Figure 7-3(b), for example, we would accomplish both guarding and reduction of ac interference by simply connecting S to either V_1 or V_2 and eliminating all ground connections.

AMPLIFIERS

Figure 7-4 shows two basic types of amplifiers. The single-ended amplifier (Figure 7-4a) grounds one side of its input and output to its case, which is normally connected to earth ground through the power cord of the recorder. The differential amplifier (Figure 7-4b) has two inputs and amplifies the difference between these inputs. The two input impedances to ground are always equal, suggesting the use of the term "balanced" or "balanced-to-ground" amplifier.

Figure 7-4(c) shows a differential amplifier which is guarded and floating. This particular configuration is becoming increasingly common as the amplifier section of recording instruments. The advantage of a differential amplifier is its ability to respond to a small difference in voltage superimposed on a larger signal. For example, we may wish to record the temperature difference between two points in a bath. Two thermocouples connected to the differential input of the amplifier will then cause the recorder to respond only to the difference in temperature. Also, any noise

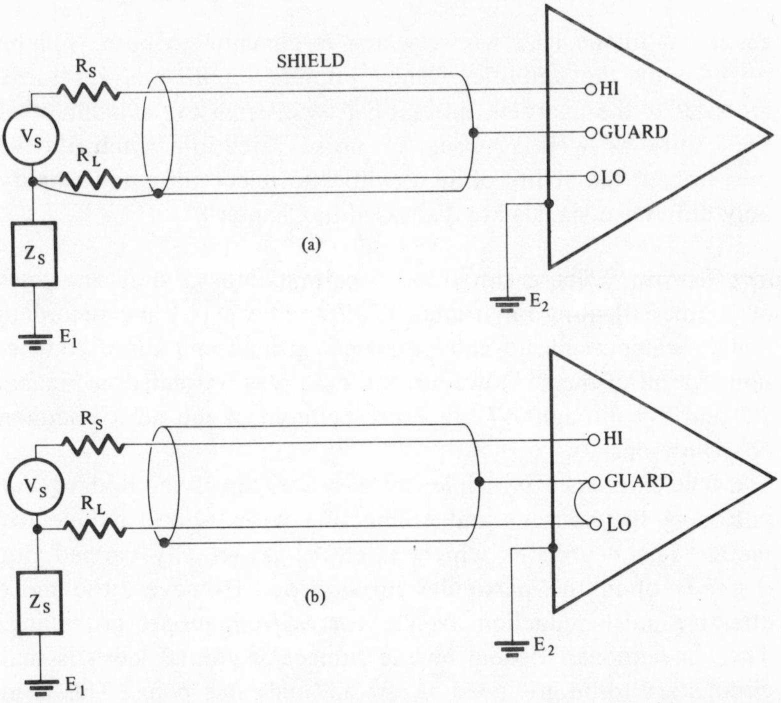

Figure 7-5. (a) The "best" solution. The signal source is floating with respect to the ground. The guard is connected to the "Lo" side of the signal *at* the signal source. Z_s is the source resistance to ground and is assumed to be large. (b) Almost as good. Although V_s is floating, there may still be some common-mode voltage which can cause some ground current through R_L.

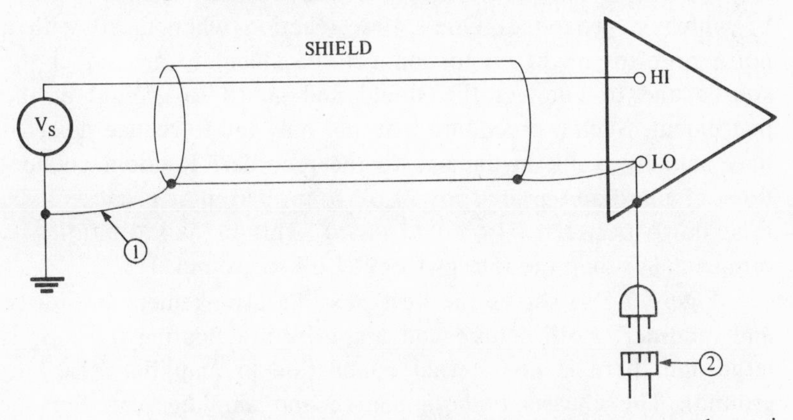

Figure 7-6. Single-ended grounded amplifier. The voltage source shown is grounded. If the voltage source is floating, connection (1) should be removed, as should the nongrounded adapter plug (2).

generated in the lead wires, which is common to both, will be rejected by the amplifier. The optimum input amplifier for a recorder is the guarded, differential type, from the standpoint of versatility as well as superior noise rejection. Quantitative measures of the ability of an amplifier to reject noise and amplify only difference signals are discussed in Chapter 8.

Applications. There are so many permutations of different types of sources (floating or grounded, differential, etc.) and recorders that it is impossible to catalog proper ground and guard connections for all of them. However, the examples presented in Figures 7-2 and 7-5 through 7-7 are representative of the most common configurations.

Shielding is, in principle, an easy task involving nothing but enclosing the sources and connecting wires within a layer of metal. The degree to which shielding is actually carried out depends upon the particular application. However, the most effective noise reduction usually comes from proper grounding. The conventional wisdom on the subject of ground loops is that circuits are to be grounded at one and only one point. This is in fact sound advice and a proper understanding of the statement together with a moderate effort at proper shielding will eliminate most problems due to noise pickup.

Figure 7-2 shows a differential amplifier and a signal source V_s which is grounded. One's first reaction when faced with a noise problem might be to connect the shield to ground at the source and to connect the shield and guard to ground at the instrument. Such a procedure may not only fail to reduce noise, it may actually result in damage to the recorder. Random connections of guard and ground, or "Lo" to ground, may cause voltage breakdown between "Lo" and guard. This breakdown rating is often smaller than the ratings from "Lo" to ground.

Figure 7-5(a) shows the best possible arrangement of source and recorder. Both source and amplifier are floating (i.e., Z_s is large and there is no internal connection of amplifier "Lo" to ground). The chassis of *both* source and amplifier can thus be grounded without causing a ground loop. This is a notable exception to the rule of only one ground in a system but it is

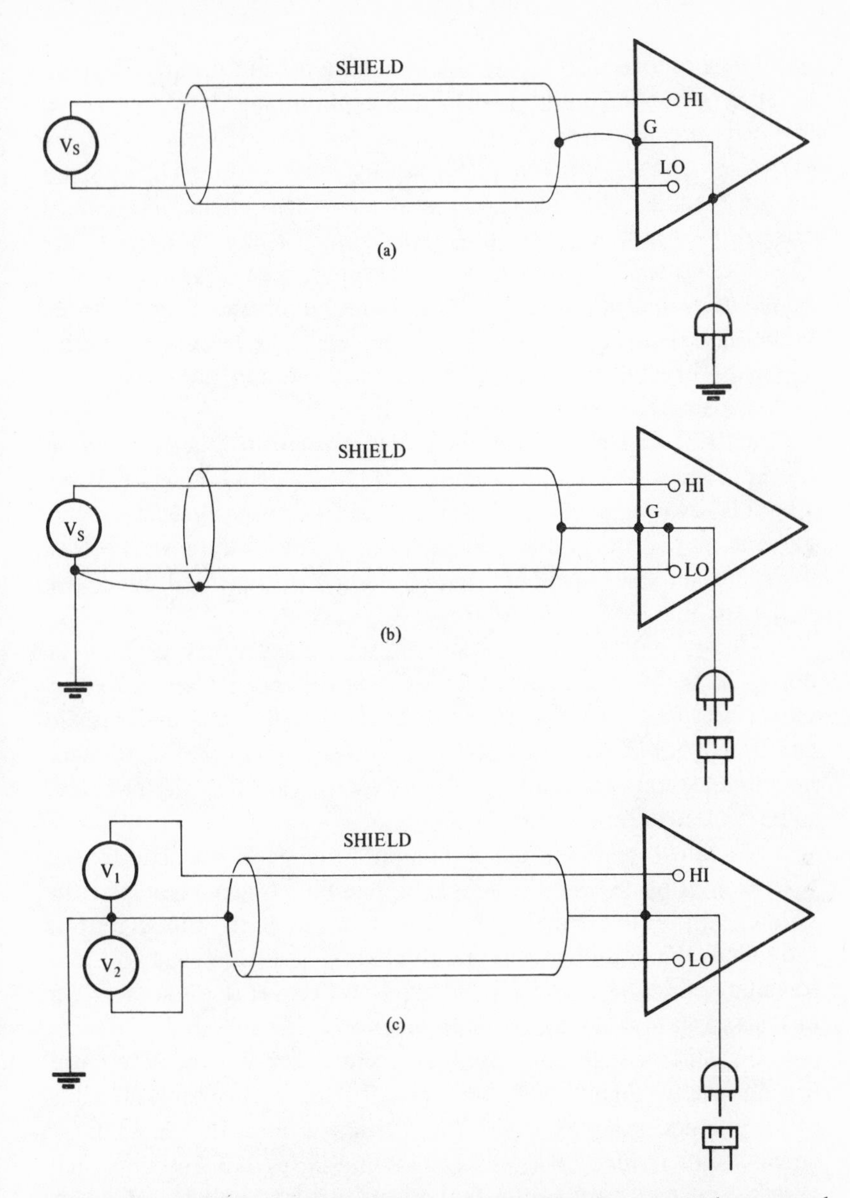

Figure 7-7. (a) The source is floating. The amplifier is connected to ground through the third prong of the power plug. (b) Single-ended, grounded source. The shield is connected to the signal "Lo," which is also ground. The amplifier "Lo" can be grounded at the amplifier, but the power cord ground of the amplifier must be removed. (c) A grounded, double-ended signal. The circuit can be grounded only at one point, so the amplifier must be floated.

possible because of the amplifier's isolation from ground. Z_s may be zero (grounded source) without degradation.

It is not necessary that the shield be grounded in order to be effective. The rejection of ac signals by the shield depends mostly on the fact that the shield is a good conductor, hence the use of copper or aluminum for this purpose. The connection of the shield or guard to circuit "Lo" serves to keep the guard at a constant potential, close to the potential of the circuit itself. Generally speaking, *the guard should not be left open (floating)*, since the breakdown voltages between guard and the circuit may be quite small (50 V).

If the guard cannot be connected directly to signal "Lo" at the source, a slight compromise can be effected (Figure 7-5b). The disadvantage is that some small common-mode (ground) current may flow through R_L, which is frequently just the lead resistance of the connecting wires. Such a current will develop a voltage which appears as an error.

When dealing with small voltages or currents, it is very desirable to have a preamplifier that is differential and guarded. However, many recorders do not have these features and we will consider a few of the more common types. When the amplifier is not floating, one must adhere rigidly to the rule that the circuit is to be grounded at only one point.

1. Single-ended, grounded amplifier (Figure 7-6). If operated with a floating signal, the amplifier should be grounded and the shield connected to low, which is ground. If the signal is grounded, the amplifier must be floated by using a "cheater" plug (disconnecting the power line ground) and preventing the amplifier chassis from making a ground connection.

2. Differential, grounded amplifier. Three typical circuits are shown in Figure 7-7. In Figure 7-7(a), it is permissible but often not necessary to connect "Lo" to ground at the amplifier since it will reduce noise levels in some cases. In Figures 7-7(b) and 7-7(c), the amplifier is floated since the signal is grounded. The signal source in Figure 7-7(c) is a grounded, double-ended signal, meaning simply that the system consists of two voltage sources, each with its own source resistance, whose common point is ground.

8

Recorder Specifications

This chapter lists all the terms commonly encountered in specifying recorder performance.

Acceleration. See "Slewing Speed."

Accuracy. Normally specified as ±%. Not to be confused with resolution or precision, accuracy is an indication of the "correctness" of any given reading. Consider an instrument with ±1% accuracy, a resolution of 0.01 V, and a linearity of ±0.01%. Suppose we have a reading of 10.08 V. The accuracy figure of ±1% indicates the true reading could be anywhere between 9.98 and 10.18 V. The *resolution* means that if our reading changes, we can distinguish between 10.09 and 10.08 V. *Linearity* tells us that if our reading changed from 10.08 to 10.09, then there was truly a change of 0.01 V. Notice that even though the true value could have been anywhere between 9.98 and 10.18 V, we could still resolve a *change* of 0.01 V.

Linearity applied to recorders relates specifically to constancy of the ratio of pen displacement to input signal. For example, if a 1 V signal gives a displacement of 51.0 cm, and 2 volts produces 102.1 cm displacement, then the linearity is given as ±0.1%.

In some cases, input impedance and source impedance will affect the accuracy of an amplifier. See "input impedance."

Amplifier Type. Balanced-to-ground, floating, differential, single-ended, etc. See Chapter 7 for an explanation of amplifier classification.

Attenuator, Variable. Essentially a variable gain control on the recorder preamplifier. For example, suppose that two adjacent scale settings on a recorder are labeled 1 V/cm and 5 V/cm; if the switch is set at 1 V/cm, the variable attenuator can make any sensitivity between 1 V/cm and 5 V/cm possible. Ideally, there should be a calibrated (fixed attenuation) and an uncalibrated (variable attenuation) control on the recorder preamp, and the scales should overlap. The advantage of variable attenuation is the ability to spread a given input signal over the entire recording.

Common Mode Rejection. Usually specified as some number of decibels (dB) at dc, another at 60 Hz, etc. A major advantage of a differential amplifier is the ability to reject signals applied simultaneously to both of its inputs. Referring to Figure 8-1, an ideal differential amplifier with a gain of A_d would have an output of

$$V_o = A_d(V_1 - V_2)$$

where $V_1 - V_2$ is the difference voltage and V_o is the output voltage. In fact, the amplifier "sees" the common voltage V_c:

$$V_c = (V_1 + V_2)/2$$

so that the real amplifier has

$$V_o = A_d(V_1 - V_2) + A_c(V_c)$$

where A_c is the gain for common mode voltages. The ratio ρ is the *common-mode rejection ratio* (CMRR)

$$\rho = A_d/A_c$$

Suppose ρ is 1000 and we have signals of $V_2 = 1.000$ V and $V_1 = 1.100$ V and we are interested in only the 0.1 V difference

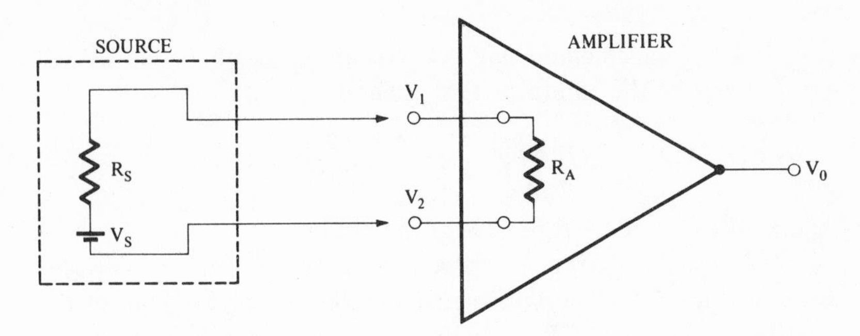

Figure 8-1. A voltage source with output voltage V_s and output (source) impedance, R_s. The amplifier has an input impedance R_A and a gain A_d.

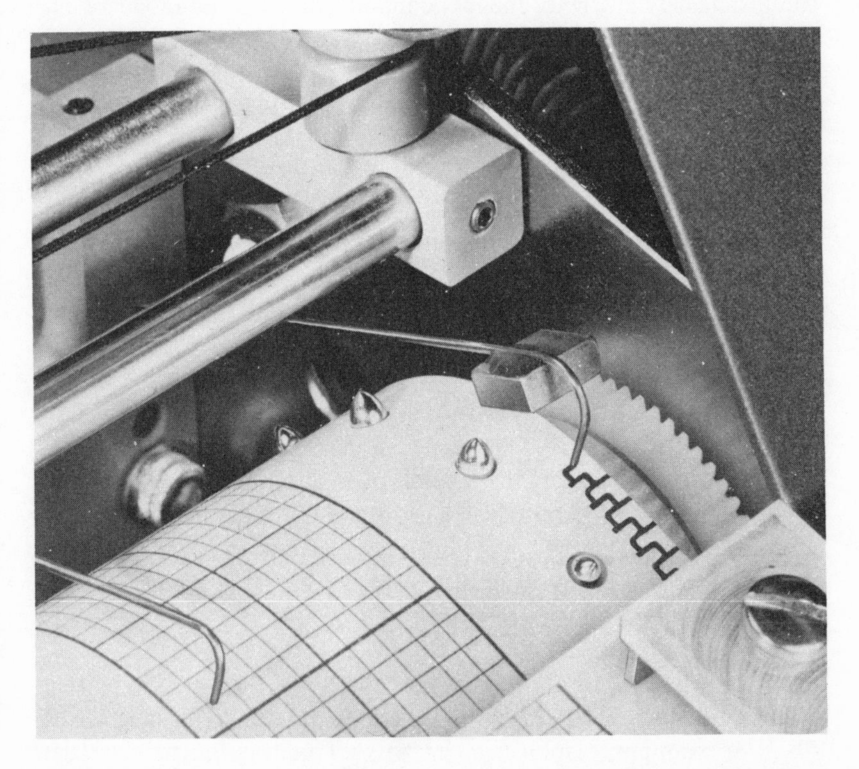

Figure 8-2. The pen on the right may be used to mark events or to make timing marks. The signal would be recorded by the left-hand pen (Sargent-Welch).

(the 1.000 V in common might be regarded as unwanted noise). Let the gain of the amplifier be A_d. Then

$$\begin{aligned} V_o &= A_d(0.1) + (A_d/1000)(1.05) \\ &= A_d(0.1) + A_d(0.001) \end{aligned}$$

The common-mode error is 1% of the reading.

Clearly, then, a high common-mode rejection ratio indicates an ability to reject spurious signals applied to both inputs of the amplifier. The CMRR is usually specified as a function of frequency because most amplifiers do not reject high-frequency common-mode signals as well as they do low-frequency or dc signals. It is possible, of course, to eliminate unwanted higher frequencies by filtering, independent of the common-mode rejection capabilities of the amplifier.

The CMRR is often stated in decibels (dB); CMRR (in dB) = 20 $\log(A_d/A_c)$. For example, a CMRR of 80 dB is the same as $\rho = 10{,}000$.

Deadband. A range of input signal for which there is no recorder response, specified either as percent full scale or a voltage. The deadband results from a combination of backlash, and friction and should be smaller than the instrument's resolution.

Event Markers. Also referred to as timers. A recorder may have an extra pen, usually located along the side of the paper. Such an "event marker" is capable of making a dot or short line in response to an external signal. If one is measuring transient response of any kind, it is often advantageous to have an event marker record the start of the process causing the response. Figure 8-2 shows an event marker.

Frequency Response. Usually specified as the maximum frequency to which the recorder responds within a certain number of dB. For example, response might be specified as flat within 1 dB (or 10%) to 100 Hz. The frequency response may be limited by the writing speed rather than the electronics. If this is the case, the maximum excursion of the trace at a given frequency may be specified. The response might be flat to 30 Hz at a 40-mm peak-to-peak deflection while being flat to 100 Hz at a 10-mm deflection.

Input Impedance. A measure of the current loading of an input signal. Figure 8-1 shows a voltage source V_s with an internal (or source) resistance R_s connected to an amplifier with an input resistance R_A. The term *impedance* is frequently substituted for resistance if the frequency response of the output is to be considered. For the simple dc case shown in Figure 8-1, the voltage that appears across the amplifier is just $[R_A/(R_s + R_A)]V_s$ and the percent error in the input is $100R_s/(R_s + R_A)$. For a 1-MΩ input impedance and a desired accuracy of 0.1%, the maximum source impedance is approximately 1 kΩ.

Recorders whose input is said to be potentiometric have a theoretically infinite input impedance since a balanced potentiometer draws no current. However, when a potentiometric recorder changes readings, there is necessarily an imbalance and the impedance presented to the source at maximum imbalance is quoted in specifications.

Chapter 7 explains that noise considerations also place practical limits on allowable source impedances. When a manufacturer states a number (10 kΩ, for example) as the maximum source impedance, it should further be specified whether the recorder will not function above this impedance (due to noise) or whether the accuracy will merely be degraded according to the discussion above.

Linearity. See Accuracy.

Paper Hold. (in recorders using separate sheets of paper). See Chapter 6.

Pen Lift. Either a manual or electronic means of moving the writing instrument away from the paper. Many instruments have the capability of remotely lifting and lowering the pen via a logic pulse or switch closure.

Reproducibility. See resettability.

Resettability. Usually specified as percent of full scale. It should be better than the percent accuracy. Resettability simply measures the ability of the recorder to give identical deflections for the same input signal.

Resolution. See accuracy.

Response Speed. Specified as time per full-scale deflection, the maximum speed at which the writing element can move. This parameter is independent of paper speed, whereas "writing speed" is not.

Slewing Speed. Essentially synonymous with response speed. Acceleration is a measure of how fast the recorder can change its slewing speed.

Stability. Usually specified as percent full scale per °C. Thermal drift is specified as an equivalent error voltage at the input. For example, 2 V/°C indicates that an equivalent error signal of 2 V will be added to (or subtracted from) the input signal for every degree Celsius change in the temperature. Some manufacturers specify drift with respect to time in an analogous manner, or percent full scale per unit time.

Time Markers. Similar to event markers. Some recorders also have internal timers to give blips of some kind at uniform time intervals so that one does not have to record or rely on the speed of the motor advancing the chart paper. (Identical in appearance to an event maker; see Figure 8-2.)

Writing Mechanisms. See Chapter 5.

Writing Speed. Specified in cm/sec or inches/sec and is a measure of how fast the pen may be permitted to move with respect to the paper.

The maximum writing speed may be limited by the highest rate at which the actual writing instrument can effectively transfer information to paper. In a photographic instrument, this would depend on the brightness of the light source and sensitivity of the photopaper. In an ink recorder, it would be affected by the pressure of the ink supply.

Zero Check. A switch to short the input terminals of a recorder to permit zero adjustment. Zero checks normally eliminate the action of the zero suppressor.

Zero Suppression (or Zero Offset). Usually expressed in terms of full scale reading (e.g., ten times full scale). This simply means that a voltage determined by the zero-suppress control is subtracted or added to the input signal. Normally this is used to offset any unwanted dc signal or to allow part of an input signal to be examined on a more sensitive scale than would otherwise be possible. The most common specification is a zero suppression equal to a full-scale reading. Quite frequently, however, it is advantageous to have an offset available equal to ten times full scale or more.

9

Troubleshooting

Except in the case of trivial malfunction, it is virtually impossible to intelligently troubleshoot a recorder without paying strict attention to the manufacturer's manual. This chapter will only mention a few common difficulties and their probable causes. Any serious work should proceed according to the steps outlined in most instrument manuals.

MAINTENANCE

Because of the presence of moving parts in a recorder, it is often the only instrument in many laboratories of electronic equipment that truly needs periodic maintenance. Slidewires, plastic strips, or wire-wound potentiometers which are contacted by the writing head as it moves need relatively frequent cleaning. Using a cleaning fluid other than one specified by the manufacturer is risky and must be done with caution. A cotton swab dipped in alcohol is usually effective in lieu of the proper fluid. Gears and other moving parts should not be lubricated unless this treatment is specifically called for in the manual. When lubrication is necessary, a high-quality instrument oil should be used. Oil should be applied sparingly, since too much oil can be as damaging as too little. If the recorder has an ink-writing mechanism, regular cleaning of the pen may eliminate some of the most frustrating disruptions. If the manufacturer offers a maintenance or pen cleaning kit as optional equipment, it is generally an excellent investment.

PERFORMANCE CHECKS

If the recorder's performance is critical, then calibration and linearity checks should be done in strict accordance with the manufacturer's instructions. For most laboratory applications, one of the circuits in Figure 9-1 will give a good indication of a recorder's performance. If a digital voltmeter is available, then the calibration of an instrument using Figure 9-1(a) is an easy matter. Linearity may be checked simply by noting displacements corresponding to 1 V, 2 V, etc. In the case of an X–Y recorder, the voltage should be applied simultaneously to both X and Y axes. Varying the potentiometer in Figure 9-1 should produce a straight line without wobble.

Figure 9-1(b) shows a calibrated voltage source that uses precision resistors and a Hg cell to produce a 1-V or 10-mV signal. If R_1 is 91 Ω and a 68-Ω resistor is put in parallel with the 270-Ω resistor (R_2), a 0.50-V signal will be provided. If the recorder is to be used where laboratory instruments are not available for calibration, a simple circuit like Figure 9-1(b) may be worth the effort to construct. Commercial units are available.

Another specification that is simple to check in a strip-chart recorder is the paper speed. The method of performing such a check is obvious and can best be done with a stopwatch. High-speed oscillographs require a different technique. Recording a sine wave from the 60-Hz line or a standard oscillator will permit convenient speed calibration.

More sophisticated features, such as noise rejection capability, acceleration, and damping, are normally performed only at the manufacturer's laboratory.

DIFFICULTIES AND THEIR CAUSES

Most common ailments afflicting recorders originate in either the mechanical (motor, gears, pen carriage, etc.) components or the

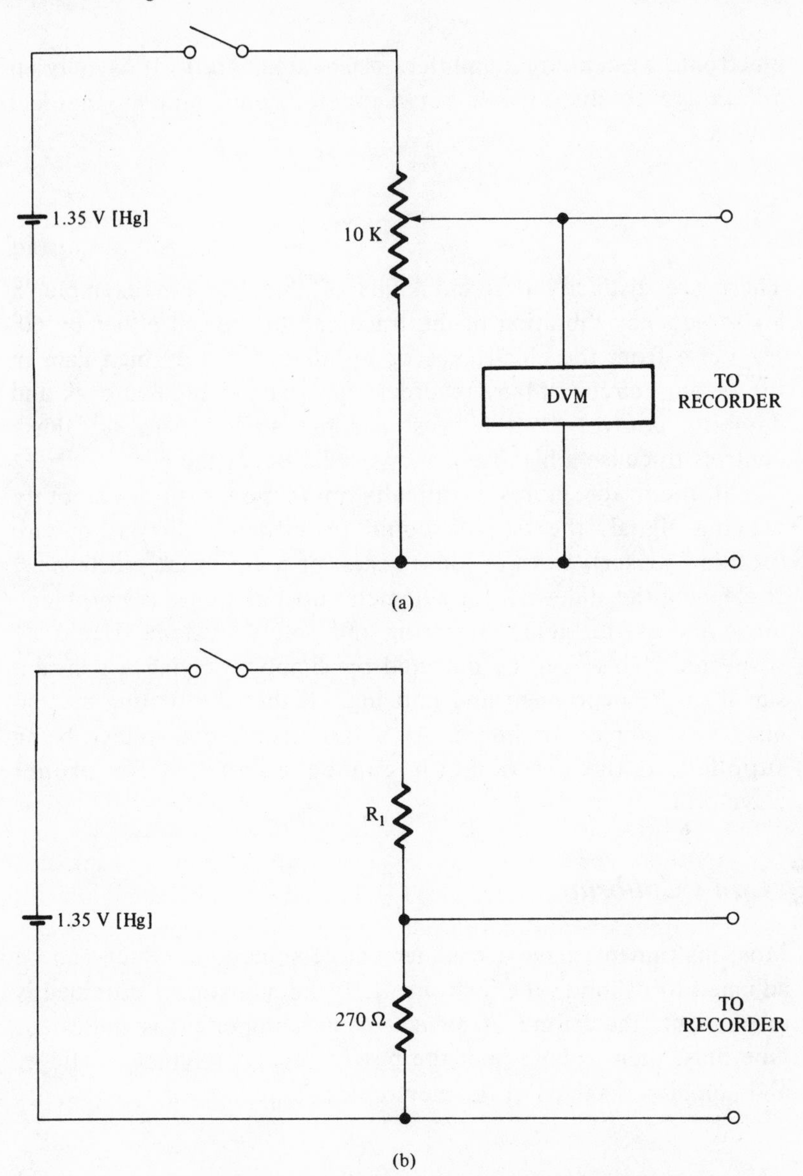

Figure 9-1. (a) A simple circuit for checking linearity and calibration. A 3½ digit voltmeter (or multimeter) is required. (b) A simple calibrated voltage source. R_1 equal to 91 Ω will give 1 V while a 3400-Ω resistor will give 10 mV. If precision resistors are used and the recorder input impedance is greater than 20 kΩ, the voltages will be accurate to better than 1%.

electronic system (preamplifier, attenuator, etc.). It is a great advantage to distinguish between electronic and mechanical troubles.

"Jittery Trace"

There are distinctly different kinds of "jitter." For example, a high-frequency vibration of the trace can be caused either by 60-Hz noise from the electronics or by an excessively high gain in the preamp circuit. Many recorders have adjustable feedback and damping controls and it is sometimes possible to set these controls to cause a high-frequency oscillation of the pen.

If the trace jumps erratically in response to a smoothly varying signal, the trouble could be either a dirty slidewire (position sensor), worn or loose gears, or noise in the electronics. If cleaning the slidewire potentiometer does not cure the problem, difficulty in the gears or string-and-pulley system should be suspected. Noise can be detected by simply applying a constant signal to the instrument and noticing whether the writing mechanism is stationary or jitters. As a last resort, the voltage being supplied to the servo motor can be monitored for proper waveform.

"Can't Calibrate"

Most instruments have a trimmer pot of some kind which can be adjusted to calibrate the instrument. If the adjustment provided is insufficient, the failure of an electronic component is indicated. One must then troubleshoot the power supply, reference voltage, and amplifier sections of the electronics.

"No Response"

If the instrument shows no activity whatever, fuses and possible loose wires should be checked. There are several possible sources of trouble. The motor or writing mechanism may be disabled or the amplifier may not be functioning. If the gears or stringing are at fault, one can usually hear the motor "trying" to respond to a signal.

The next step is to locate the output of the amplifier at the point where it connects to the motor. The proper signal may be either ac or dc, depending on the recorder, but the magnitude of the signal should change in response to an input signal. No signal at this point calls for a check of the amplifier after making sure the motor is not shorted and causing the loss of signal.

If the motor seems to be at fault, one should first check the resistance of the windings. A "burned-out" motor will frequently show an open circuit. In this case, the motor needs to be replaced.

If the windings are intact, further tests can be made. If the motor is a dc type, it will probably have only two wires coming from it. A sufficient dc voltage applied to the wires should cause the motor to turn. More often, the motor will be of a two-phase ac type, with four wires emerging. Generally speaking, two of the wires will go to the ac line to provide the power to the motor. Two other wires will come from the amplifier and provide control of the motor speed. The voltage required by the control windings is generally less (35 V, peak) than the power winding voltage (full line voltage). Occasionally, the capacitor whose function is to change the phase of the voltage seen by the motor windings will fail. This is hard to detect since the motor can still receive the correct line voltage but with an improper phase, causing no rotation or rotation when there should be none.

"Partial Response"

If there is no signal response but the zero offset adjustment still works, the trouble is then localized to the input preamplifier. If the response is sluggish, the fault may be due to worn out batteries in the balance circuit or insufficient gain in the amplifier.

"Writing Mechanism and Restringing"

Occasionally, a string or spring in the mechanical system will break or come loose. Restringing a recorder is often a major undertaking (Figures 3-7, 4-3, 4-4) and a decision must be made about returning the recorder to a repair center or attempting the restringing oneself. Restringing instructions are usually in the recorder manual.

10

Recorder Accessories

There is an abundance of auxiliary devices available to increase the usefulness and convenience of recorders. Some of these must be installed within the recorder case itself, while others are intended for external use. A selection of such accessories is described in this chapter.

INTEGRATORS

For many applications, it is desirable to be able to integrate continuously the area enclosed beneath a curve. This can be accomplished automatically either by electronic means or by a mechanical device. Conventionally, in general-purpose strip-chart recorders, the integrator provides a signal to an auxiliary pen, in the form of a sequence of pulses. The integrator controls the frequency of the pulse train, and the value of the integral is read by counting the pulses recorded by the added pen. In some recorders, notably those built-in to magnetic resonance (NMR) spectrometers, the integrator traces a step curve on the same area of chart with the principal spectrum. Figure 10-1 shows examples of both modes of presentation.

There are a number of electronic methods, both analog and digital, of integrating a signal as a function of time. One widely used circuit is a high-gain (operational) amplifier with capacitive feedback. The amplifier forces the capacitor to charge at a rate determined by the magnitude and sign of the signal voltage; the potential developed across the capacitor is then the time-integral of the signal. This device is excellent for integration times up to a minute or so, but requires premium components for longer times if high precision is sought. The output most conveniently gives the step type of presentation shown in Figure 10-1(a).

Another electronic approach is the voltage-to-frequency converter. An example is shown in Figure 10-2. The amplifier forces a current proportional to the signal voltage to pass through a light-emitting diode (LED) optically coupled to a photoresistive cell. The latter is part of a pulse generating circuit utilizing a unijunction transistor (UJT). The rate of production of pulses is thus determined by the signal level. The integral is recorded as in Figure 10-1(b).

A mechanical integrator often employed is based on the ball-and-disk mechanism depicted in Figure 10-3. The device consists of a rotating disk in friction contact with a ball, which in turn makes contact with a roller. The roller rotates as the result of power transferred to it by the ball and disk. The ball is held in position by a retaining frame which can move radially across the disk. This linear motion is taken from the pen carriage of the recorder, and the rotation of the disk is synchronous with the paper feed. Mathematical analysis shows that the rotation of the roller is proportional to the area beneath the curve being traced on the recorder.

DIFFERENTIATORS

It is sometimes advantageous to be able to plot the slope of a function as well as the function itself. This will assist, for example, in the resolution of partially overlapping features. Differentiators are, however, less commonly supplied than integrators. The time derivative of a signal can be sensed by a variety

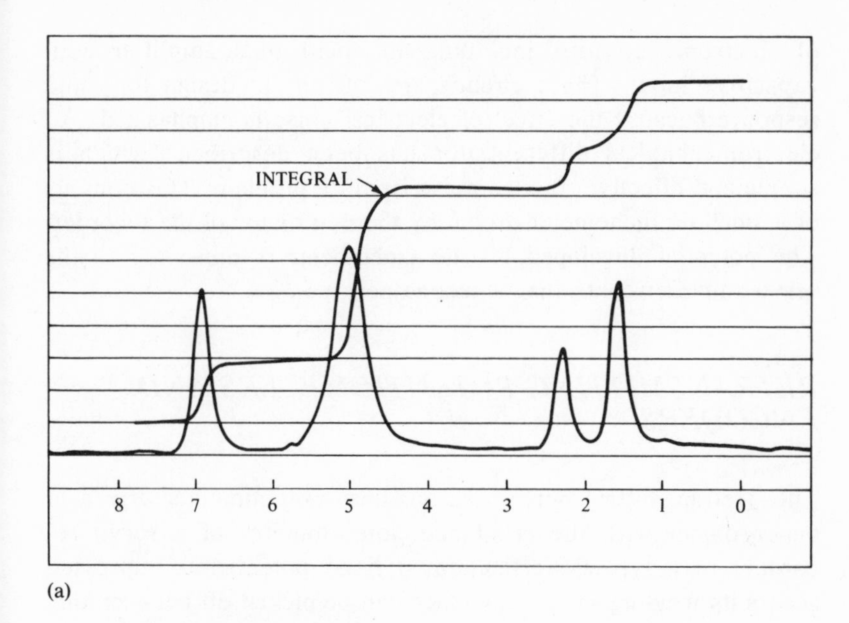

(a)

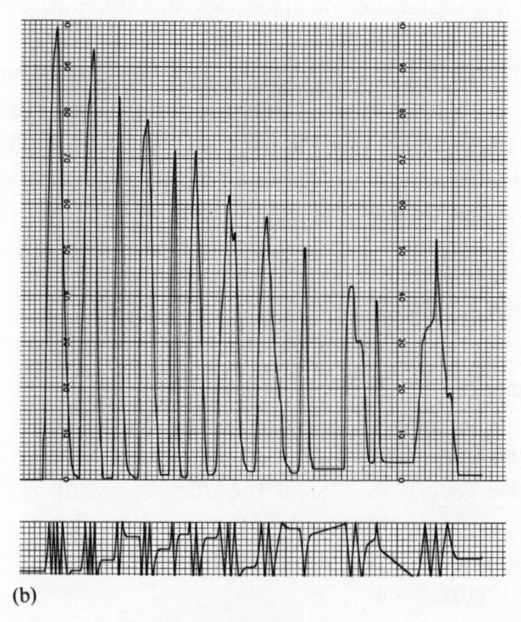

(b)

Figure 10-1. (a) The trace labeled Integral gives the area under the other curve. (b) The number of cycles of the lower trace gives the area under the upper curve.

of electronic circuits, including an operational amplifier with capacitive input. These circuits are difficult to design for rapid response because the effect of electrical noise is emphasized. An electromechanical differentiator has been described* which is simple and effective, though noise is still a problem. This consists of a small dc tachometer driven by the pen motor of the recorder. The potential developed by the tachometer is presented to the servo amplifier controlling a second pen.

RETRANSMITTING POTENTIOMETERS AND ENCODERS

The "retransmitting pot" is an auxiliary potentiometer driven in synchronism with the rebalance potentiometer of a servo recorder. In a typical application, a fixed potential is impressed across its winding so that a voltage can be picked off between one end and the variable tap, proportional to the signal being recorded. This signal can be used, for example, to drive a slave recorder.

A shaft encoder is a device, also driven synchronously with the rebalance potentiometer, to give a pulse-coded signal in BCD (binary-coded decimal) or other similar format, equivalent to the magnitude being recorded. This can be used to operate a digital printer, or to interface with a digital computer.

EVENT MARKERS

An event marker is an extra pen that can be installed in many strip-chart recorders to make a distinctive "blip" along the margin at the command of an external switch. It can be used to give a precise record in time of the occurrence of any significant event. With a spectrum recorder, for instance, the passage of the scan through each integral wavelength could be recorded automatically. (See Figure 8-2.)

* E. C. Olson and C. D. Alway, *Anal. Chem.* **32**, 370 (1960).

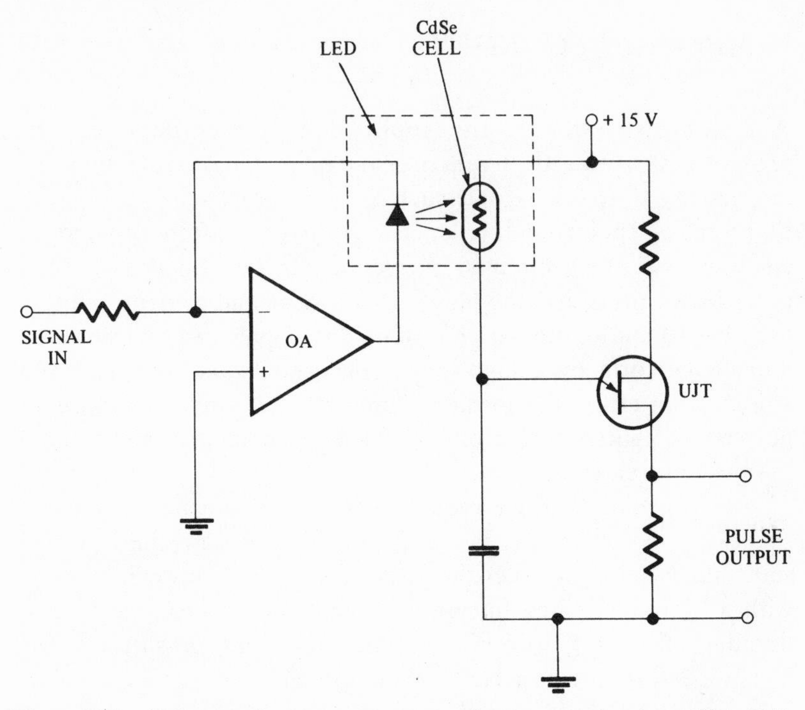

Figure 10-2. Voltage-to-frequency converter using an operational amplifier (OA), an optically coupled isolater (LED and CdSe photoresistive cell), and a unijunction (UJT) pulse generator.

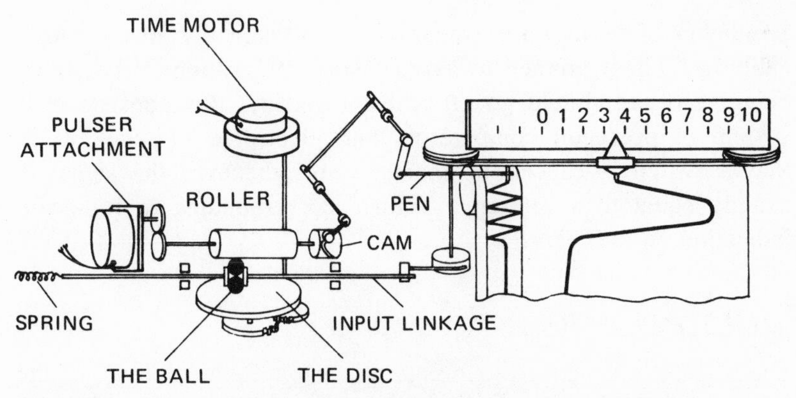

Figure 10-3. Principle of the ball-and-disk integrator. The second (idler) ball is not essential, but merely a design convenience. The rotation of the roller is sensed by a "spiral-in, spiral-out" cam and actuates an auxiliary pen to record the number of revolutions. (Disc Instruments.)

SAMPLING RECORDERS

A major restriction in the application of recorders is their relatively slow speed. Even oscillographic types are limited in their ability to follow signals such as those produced in a time-of-flight mass spectrometer, or the transients often present in electrical switching circuits. A record can be obtained in such cases by means of a cathode-ray oscilloscope and camera, but this is awkward and of limited precision. The signals can be translated into digital form by a high-speed analog-to-digital converter and stored in an electronic memory bank for later study; the special-purpose computers for accomplishing this are beyond the scope of the present treatment.

Fast signals that are repetitive in nature can be sampled at successive points in successive cycles, and the samplings plotted sequentially to reconstruct the actual waveform. This can be done with a "boxcar" gated integrator, commercially available from a number of manufacturers. A simplified version suitable for construction by the user has been described.*

SCALE EXPANDERS

A number of devices are available for automatically introducing a calibrated offset voltage to avoid losing information if the signal being recorded should go off-scale. Typically, this consists of a system of precision comparators and electronic switches, such that a switch is caused to change state whenever the signal is exactly equal to a reference voltage corresponding to full-scale indication on the recorder.

MULTIPLEXERS

One recorder can be made to do the work of several by electronic switching (multiplexing) between several inputs. Figure 10-4

* T. I. Smith, *Rev. Sci. Instrum.*, **44**, 288 (1973).

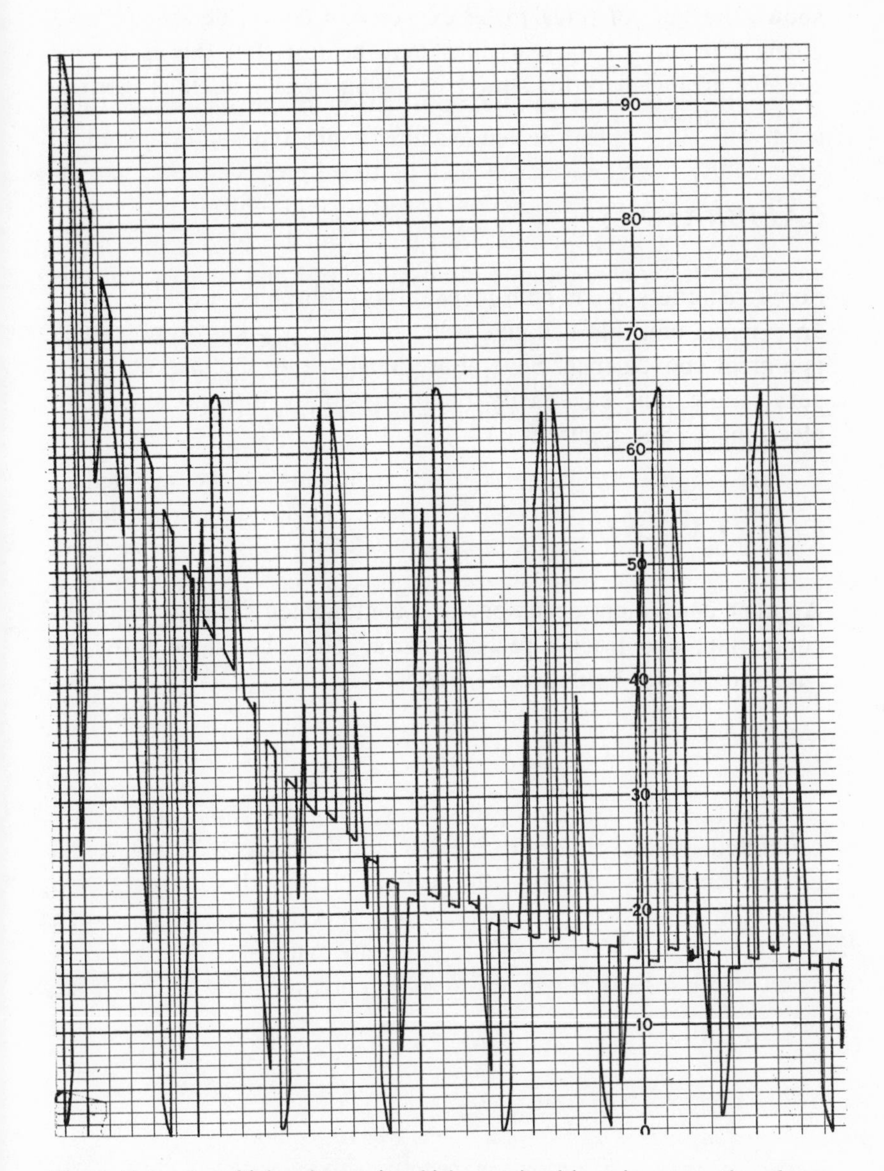

Figure 10-4. A multiplexed trace in which one signal is a sine wave, the other a decaying exponential (Donlee Laboratories, Inc.).

shows the type of trace to be expected. Clearly, details of rapid changes in any of the variables may be lost, but this is a very suitable arrangement for slowly changing signals, such as thermocouple voltages.

CALIBRATORS

Some recorders have built-in calibration circuits. Lacking such, an external unit that will give selected, precisely known potentials is a great convenience. Several are available on the market, and a serviceable unit can easily be designed by anyone skilled in electronics. (See Figure 9-1.)

CHART VIEWERS

Another convenient accessory is a back-lighted table for reading, comparing, and tracing record charts. This should be provided with spindles for holding paper rolls.

Appendix: Sources of Laboratory Recorders

Amprobe Instrument
630 Merrick Road
Lynbrook, New York 11563

Astro–Med
Atlan–Tol Industrial Park
West Warwick, Rhode Island 02893

B & F Instruments
Cornwells Heights
Pennsylvania 19020

Barber–Colman Company
Industrial Instruments Division
1300 Rock Street
Rockford, Illinois 61101

Beckman Instruments
2500 Harbor Boulevard
Fullerton, California 92634

Bell & Howell, Control Products Division
706 Bostwick Avenue
Bridgeport, Connecticut 06605

Brinkmann Instruments
Cantiague Road
Westbury, New York 11590

Bristol Division of ACCO
40 Bristol Street
Waterbury, Connecticut 06720

Carle Instruments, Inc.
1141 E. Ash Avenue
Fullerton, California 92631

Computer Instruments Corp.
92 Madison Avenue
Hempstead, New York 11550

Dohrmann Division of Envirotech
3240 Scott Boulevard
Santa Clara, California 95050

Enraf–Nonius, Inc.
130 County Courthouse Road
Garden City Park, New York 11040

Esterline Angus
P.O. Box 24000
Indianapolis, Indiana 46224

Fisher Scientific Company
711 Forbes Avenue
Pittsburgh, Pennsylvania 15219

Gamma Scientific, Inc.
3777 Ruffin Road
San Diego, California 92123

General Scanning, Inc.
150 Coolidge Avenue
Watertown, Massachusetts 02172

Gould, Inc, Instrument Systems Division
3631 Perkins Avenue
Cleveland, Ohio 44114

Gulton Measurements and Controls Division
Routez and Middle Road
East Greenwich, Rhode Island 02818

Hathaway Industries, Inc.
11616 East 51st Street
Tulsa, Oklahoma 74145

Heath–Schlumberger Instruments
Benton Harbor
Michigan 49022

Hewlett–Packard Company
1501 Page Mill Road
Palo Alto, California 94304

Hewlett–Packard Company, San Diego Division
16399 West Bernardo Drive
San Diego, California 92127

Honeywell Inc., Process Control Division
1100 Virginia Drive
Fort Washington, Pennsylvania 19034

Honeywell, Inc., Test Instruments Division
P.O. Box 5227, 4800 E. Dry Creek Road
Denver, Colorado 80217

Houston Instrument Co., Division of Bausch & Lomb, Inc.
8500 Cameron Road
Austin, Texas 78753

Lab-Line Instruments
15th and Bloomingdale
Melrose Park, Illinois 60160

Lafayette Instrument Company
P.O. Box 1279, Sagamore Parkway
Lafayette, Indiana 47902

Leeds & Northrup Company
Sumneytown Pike
North Wales, Pennsylvania 19454

Linear Instruments Corp.
17282 Eastman
Irvine, California 92705

LKB Instruments, Inc.
12221 Parklawn Drive
Rockville, Maryland 20852

McKee–Pedersen Instruments
P.O. Box 322
Danville, California 94526

MFE Corporation
Keewaydin Drive
Salem, New Hampshire 03079

Philips Test and Measuring Instruments, Inc.
400 Crossways Park Drive
Woodbury, New Jersey 11797

Photovolt Corporation
1115 Broadway
New York, New York 10010

San–Ei Instrument Co., Ltd.
1-57, Tenjin-cho, Kodaira City
Tokyo, Japan

Sargent–Welch Scientific Co.
7300 N. Linder Avenue
Skokie, Illinois 60076

Siemens Corporation
186 Wood Avenue South
Iselin, New Jersey 08830

Simpson Electric Company
853 Dundee Avenue
Elgin, Illinois 60120

Soltec Corporation
10747 Chandler Boulevard
North Holywood, California 91601

Texas Instruments, Inc.
P.O. Box 1444
Houston, Texas 77001

Varian Aerograph
2700 Mitchell Drive
Walnut Creek, California 94598

Varian Instrument Division and Varian Data Machines
611 Hansen Way
Palo Alto, California 94303

Westronics, Inc.
5050 Mark IV Parkway
Fort Worth, Texas 76106

Appendix Table: Sources

Manufacturer	Deflection recorders					Nulling recorders							Rebalance device		Pen motor			
	DC	AC	Oscillographic	Multichannel	Other	Potentiometric	R (or Z) bridge	Multichannel[b]	Multichannel[c]	Zero offset	Multiple range	Other	Potentiometric	Other	Rotating <1 turn	Rotating, multiturn	Linear motion	Other
Amprobe	•	•																
Astro–Med	•	•	•	•											•			
B & F			•															
Barber–Colman						•	•		•	•	•					•		
Bausch & Lomb[d]																		
Beckman Instruments	•	•	•	•									•					
Bell & Howell													•		•			
Brinkmann Inst.						•	•		•	•	•	•	•			•		
Bristol Div. (ACCO)						•			•	•		6	•					
Carle						•												
Computer Instruments																		
Enraf–Nonius						•			•	•	•		•		•	•		
Envirotech																	•	
Esterline Angus	•	•		•	17	•		•	•	•	•					•	•	
Fisher						•							•					
Gamma Scientific																		
General Scanning	•	•	•	•											•			
Gould	•	•	•	•													•	
Gulton	•	•	•	•	24	•		•		•	•				•	•	•	
Hathaway			•	•														
Heath–Schulumberger						•				•	•							
Hewlett-Packard (*P.A.*)	•	•	•	•														
Hewlett-Packard (*S.D.*)	•	•	•	•		•	•			•	•		•					
Honeywell Proc.	•		•			•			•	•	•	•	•	•		•		
Honeywell, Test	•	•	•	•	31													
Houston	•	•		•		•			•	•			•	36				
Lab–Line											•							
Lafayette	•	•		•		•	•	•	•	•	•		•		•		•	
Leeds & Northrup						•	•	•	•	•	•							
Linear						•		•	•	•	•							
LKB	•	•		•		•			•		•							
McKee–Pedersen						•				•	•					•		
MFE Corp.	•	•	•	•		•		•	•	•	•		•		•			
Philips						•	•	•	•	•	•		•					
Photovolt						•							•			•		
San–Ei			•			•									•		•	
Sargent–Welch						•			•	•	•				•	•		
Siemens			•	•		•		•	•	•	•		•				•	
Simpson Electric						•		•		•	•		•					
Soltec	•	•	•	•		•		•	•	•	•					•		
Texas Inst.	•	•		•		•	•	•	•	•	•					•	•	
Varian Aerograph						•		•	•	•						•	•	
Varian Instrument	•					•		•	•		•		•			•		
Westronics						•		•	•	•	•	60						

[a]See legend on following page.
[b]Side-by-side.
[c]Overlapping.
[d]See Houston Inst.

of Laboratory Recorders[a]

X-Y recorders			Paper				Writing method									Paper feed					Speed change			Accessories available				
Plug-in functions	Time-base	Other	Continuous rolls	Fan (Z) fold	Discrete sheets	Other	Capillary pen	Pressurized ink	UV sensitive	Heat-sensitive	Pressure-sensitive	Xerographic	Electric	Fixed styli	Other	Synchronous motor	Spring motor	Reversible servo	Stepping motor	Other	Gears	Other mechanism	Electronic	Mechanical integrator	Electronic integrator	Differentiator	Event marker	Other
			•								•					•						•	•				•	
			•	•						•			•			•				1	•							2
																											•	
			•				•								3	•					•	4						5
							•			•	•					•			•		•		•	•	•		•	
			•				•									•					•	•	•	•	•		•	•
	•														•	•					•			•			•	
•		•	•			7	•								8	•		•	•	9	•	10			•			11
			•				•									•												
																												12
•	•		•		•		•								13			•	•				•		•		•	11
						14	•									•			•		•	15	•	•			•	16
•	•	18	•		•		•			•						•	•		•		•	19	•		•		•	
			•												20				•			21			•		•	
		22					•													23							•	
			•	•						•						•					•							
•	•		•	•				•		•						•					•				•	•	•	
			•	•						•	•		•			•			•		•	25	•				•	26
									•			•	•										•				•	
•			•												27				•									28
•	•	•	•	•	•		•	•		•																	•	
•	•	19					•	•		•			•			•					•	29	•	•	•	•	•	30
			•		•		•			•						•			•		•				•		•	
			•						•						32	•		•		33	•	34	•				•	35
•	•		•	•	•		•								37	•		•	•				•		•		•	38
			•							•						•					•							
			•	•			•			•						•					•		•	•	•		•	•
•	•		•	•			•			•	•				39	•		•	•		•	40						41
			•				•									•			•				•		•	•	•	
			•				•				•			•		•					•		•		•		•	42
		43	•												44				•				•					
•	•		•	•	•		•			•					•	•			•			45	•		•		•	46
•	•		•	•	•		•	•								•		•	•									47
	•		•				•									•					•	48			•		•	
			•	•			•		•		•		•			•					•	•			•		•	
			•				•								49	•			•			50	•				•	51
											•									52	•	•		•			•	•
•	•		•				•	•			•	•		•		•	•	•			•		•					
•	•		•	•	•		•						•			•	•		•		•	53	•				•	
•			•			54	•			•	•			•	8	•		•	•	55	•	56	•	•	•		•	57
			•	•			•							•					•			•	•	•			•	
		58	•	•			•								63				•				•	•			•	59
			•				•				•				61	•			•		•	62	•	•			•	63

Legend for Appendix Table

1. Variable DC
2. Timers, chart take-up
3. Metal print wheel
4. Dual- and eight-speed transmission
5. Control switches, limit switches, three mode controllers
6. Multipoint; also recorders with plug-in range card for quick range changes
7. Round charts
8. Disposable cartridge (fiber tip)
9. Selsyn drive system
10. Electric and mechanical (turret changer)
11. Retransmitting slidewire alarm contacts
12. Potentiometric resistance elements (rotary, linear motion, arc segments)
13. Fiber tip pen
14. Fits Dohrmann recorder only
15. Stepping motor
16. High and low level alarms
17. Pressure, position, vacuum
18. XYY′
19. Pulse change
20. Disposable nylon- or Dacron-tipped pens
21. Push-button
22. Low-cost, general
23. Fixed
24. Plug-in signal conditioners
25. Electrical multiple-speed transmission
26. Alpha-numeric printout
27. Disposable nylon-tip pen
28. Centimeter chart drive, low pass noise filter, limit detector module, multiplexer module, DC offset module, potientiometric amplifier module
29. Electrical
30. Metric, English, chart, speeds, etc.
31. Linescan
32. Magnetic styli, fiber-optic cathode ray tube
33. Wide range DC servo
34. Servo
35. Time lines, grid lines, multiplexor signal conditioners for electrical and transducer signals
36. Capacitance
37. Ball point
38. High-speed servos, point plotters, log converters
39. Disposable cartridge
40. Dual synchronous motors, reversible
41. Log amplifier, pen lifter
42. Magnetic pen drop
43. Optional plug-in time base
44. Acrylic fiber-tip pen
45. Electro-mechanical
46. Alarms circuit, retransmitting buffer amplifier, remote chart drive, proportional remote chart drive, lock knobs, electronic remote pen lift
47. Thermocouple range modules, event marker, retransmitting potentiometric, rack mounts, time bases
48. Change motor
49. Ball point and nylon-tip pen
50. Motor change
51. Log-linear converter, input attenuator, upper-lower limit controller
52. DC
53. Switch-selected
54. Continuous roll with perforations
55. Synchro
56. Transmissions
57. URAP and USAP units, variable input ranges, variable zero control, retransmit pots., T/C inputs, RTD inputs, pen lifter, doors, lighting kits, platens
58. Electrostatic
59. Alarm contacts, remote start/stop, remote chart drive
60. Multipoint
61. Ball point, fiber tip
62. Dual motor
63. Alarm switches, retransmitting slidewire

Index